Betting Against the Crowd

Yair Neuman

Betting Against the Crowd

A Complex Systems Approach

Springer

Yair Neuman
Department of Cognitive and Brain Sciences
Ben-Gurion University
Beersheba, Israel

ISBN 978-3-031-52018-1 ISBN 978-3-031-52019-8 (eBook)
https://doi.org/10.1007/978-3-031-52019-8

This Springer imprint is published by the registered company Springer Nature Switzerland AG
The registered company address is: Gewerbestrasse 11, 6330 Cham, Switzerland

Paper in this product is recyclable.

This book is dedicated to my friend S. Gozlan, who played twice with Lady Fortuna and won.

Preface

Knowing is not understanding. There is a great difference between knowing and understanding: you can know a lot about something and not really understand it.
—*Charles F. Kettering*

There is an unexplained and shocking gap between our knowledge and our understanding. We all know this, but we are surprised each time we realize it, an experience that recursively supports the existence of this gap. I am no exception. When Netanyahu's government was established, I knew its components were not the bread and butter of a democratic government. However, it came to me as a total surprise that the coalition of these components is actually striving to destroy Israel's democracy at any cost.

In a deep sense, my surprise was similar to what people feel when they unexpectedly come under a violent attack. We know these things happen, but we don't get any deep understanding until we experience them at first hand. In his book, Miller (2008, p. 55) describes four basic truths of violent assault: "Assaults happen closer, faster, more suddenly, and with more power than most people believe." Most people are not trained to handle this kind of violence. They may be talented martial "artists," big athletic guys, or just the tough guys in their high school, but as Miller explains, they are not prepared for this kind of attack, and may therefore be surprised and shocked to the extent of being totally destroyed.

While I feel unable to fully determine my motivations for writing the present book, I am convinced that the situation in Israel was an important trigger. Moreover, in October 2023, the radical Islamists of Hamas launched the most devastating terror attack Israel has ever experienced. Before the attack, I completed a research paper analyzing the Palestinian–Israeli conflict and explaining how poorly prepared we were for a bombshell of these proportions.

Being hit so hard and so suddenly teaches us another important lesson. This book draws on my scientific work and presents an attempt to understand the dynamics of crowds. Through this scientific understanding, it aims to identify the individual's place within the collective and point out ways she can bet against the crowd. Asserting one's individuality within the crowd is possible, but as will be explained, it is a never-ending challenge. For such a challenge, it is better to be well prepared.

This is where it is useful to build up a toolkit of ideas and lessons gained by studying non-linear and complex systems. This book provides just such a toolkit and applies it in different contexts, from politics to sport and finance. The reader won't find recipes for betting against the crowd, but rather a toolkit of ideas illustrated through experiment, theory, common sense, and humor.

Beersheba, Israel Yair Neuman

Reference

Miller, R. (2008). Meditations on violence. USA: YMAA Publication Center.

Acknowledgments I thank Yochai Cohen for his indispensable support in my work. I also thank Grzegorz Wilk for commenting on previous papers and my editor, Angela Lahee, for her support and trust in my academic ventures.

Summary

Crowds are misleading in both their simplicity and their complexity. On the one hand, they behave according to expected trends, and on the other hand, they present sudden shifts and frantic, unexpected behavior. Therefore, "betting against the crowd," whether in politics, sports, or finance, requires a deep understanding of crowd dynamics. In this book, Prof. Neuman addresses this challenge by delving into the complexity of crowds. The book exposes foundational issues and presents novel ideas, such as why our understanding of crowds decays exponentially, how to use short-term prediction to bet against the crowd in financial markets, and why the long tail of fatalities in armed conflicts leaves us surprised by the blitz attack of violent mobs. The book combines scientific knowledge, experiments, and friendly, humoristic exposition that will interest anyone who seeks to understand crowds and sometimes wishes to act within and against them.

Contents

Part I

Foundations of Crowd's Dynamics

1

Navigating the Collective: Insights into Crowd Behavior and Strategies for the Individual

From the Painted Bird to the Celebrating Crowd

> … an agglomeration of men presents new characteristics very different from those of the individuals composing it. (Bon, 1895, p. 2)

Human beings group into various "agglomerations," from the coalition of Iranian women struggling against the Ayatollahs' oppression to the mob of football hooligans violating public order and the crowd celebrating the carnival in Rio. Understanding the behavior of these agglomerates is an old challenge, and the mind of the collective has been expressed and studied in numerous scientific and artistic works. For example, "The Painted Bird" (Kosiński, 1965) is a novel that provides a powerful artistic description of the exclusion and painful destiny of social outsiders, and Bakhtin (1984) gave us important insights into the mind of the collective celebrating the carnival. It would seem that nothing new could be added to this vast literature. However, as Le Bon (1895) observed in his classic, the agglomeration of people is a whole *different* from the sum of its parts. A group may be composed of good citizens. Each and every individual may be of good character, but when put together under the leadership of a charismatic person, they may turn into a murderous mob. One may then doubt the human quality of these good Samaritans, arguing that each and every individual comprising the murderous mob must actually be a hidden murderer and that our failure to understand the emerging behavior of the mob is a failure to understand the dark side of

Y. Neuman, *Betting Against the Crowd*, https://doi.org/10.1007/978-3-031-52019-8_1

its components. For the outside observer, like Le Bon, what is inside any individual's mind is less important when observing the crowd's behavior. Most of those making up the mob just mentioned may never have carried out a deadly deed. But within the mob, they may do so. To understand this point, watch *Dogville*[1] by Lars von Trier is a tantalizing film showing how ultimate evil emerges from a collective of good American citizens. Watch it if you have your own doubts about the complexity of social wholes and the evil that human beings can impose on their fellow men and women.

In this context of a whole different from the sum of its parts, we are in the realm of non-linear systems, where uncertainty and surprise may have the upper hand. For instance, imagine two different societies of equal size. One is a collective of hunter–gatherers where the distribution of "wealth" is almost equal. The other is one where the distribution of wealth is unequal, in fact, a Pareto-style distribution where 10% of the population holds 90% of the wealth. What happens if, through some malicious experiment, we mix the two populations? What would the new society look like in terms of its wealth distribution? A naïve hypothesis might be that the new distribution of wealth will simply be the sum of the two previous distributions. This hypothesis is grounded on the assumption of *additivity*. However, we would not scientifically bet on this result for a simple reason: *interactions*. As nicely explained by Rovelli (2019), we are biased to think about the "essence" of things, where sometimes the most important thing to look at is change and dynamics taking place through interactions.

An *event*, such as a kiss, explains Rovelli, is not a thing. It does not persist in time and space like my kitchen table. The vicious behavior of Dogville's citizens is grounded in group dynamics. In interactions. Some information exists *in-between* the interacting components (Neuman, 2021), and this information cannot be reduced to the psychology of each and every individual who is a part of this "wonderful" community of good Americans. However, sometimes, a reduction is possible. A group of notorious South American gang members is probably composed of violent individuals with violent group behavior. As we can see, the agglomerate discussed by Le Bon may come in different forms.

[1] https://en.wikipedia.org/wiki/dogville.

The Importance of Interactions

The interactions between "human particles" are not the same as those between gas particles. In fact, all living systems, from the cell to the snail, exhibit the miracle of emerging behaviors irreducible to the simple behavior or sum of their components. Something miraculous happens when things interact and produce, on a higher level of analysis, a behavior that cannot be trivially explained by reduction to the lowest level of the aggregate; this miracle also holds for human collectives. Simple explanations of crowd behavior sometimes ignore this fact realized by Le Bon long ago. For example, Rosner and Ritchie (2018) present the results of their optimism opinion survey covering 26,489 people across 28 countries.[2] Their results are presented in terms of simple percentages. For example, 41% of the Chinese respondents "think the world is getting better." This result is significantly higher than the 3% of French respondents who think the same. These results might be taken to imply that Chinese optimism is 14 times greater (!) than French optimism. Chinese and French optimism seem to be on a different scale. For comparison, the ratio between Chinese and French optimism is almost the same as the ratio between the heights of the Eiffel Tower and a giraffe. We may imagine 41% of the current 1,425,572,821 people living today in China starting their day with a smile, reminding themselves that the future under the eternal leadership of Xi Jinping is more promising than ever, while 97% of the French begin their day by dipping a butter-saturated croissant in coffee and gazing depressively into the future.

Is it the case that the Chinese collective is more optimistic than the French one? And if so, in what sense, except for the trivial sense that a greater portion of the Chinese respondents answered the optimism question positively? There is a fact: 41% of the specific sample of respondents agreed with the specific optimism item. However, jumping to the conclusion that the Chinese collective, whatever it is, is optimistic requires a leap of inference. The Chinese collective may *behave* differently from the sum, or some of the sum, of individuals comprising it. This lesson has not been sufficiently built into our understanding, although it has been pointed out by people from Le Bon to Bateson (Bateson, 2000).

So far, I have emphasized two points. First, a collective of human beings is a whole different from the sum of its parts. Second, it is the outcome of interactions. The first point urges us to examine the behavior of collectives by avoiding the naïve assumption that they are the sum of their parts

[2] https://www.ipsos.com/en/global-perceptions-development-progress-perils-perceptions-research.

or that their behavior scales linearly with the size of the system. Crowds do not behave like a single individual multiplied by 10,000. The second point urges us to examine the collective as an event constituted through micro-level interactions, with possibly surprising and unexpected results. Again, this understanding is important to avoid poor explanations which merely attribute some kind of "essence" to the collective. Le Bon made this mistake by discussing race as an innate explanation of a crowd's behavior. Today, with the exception of anti-scientific and zealous racists, this essentialist explanation had been rejected. Therefore, we are left with complex wholes and the challenge of understanding them while maintaining a delicate balance between authentically representing their complexity and our need to simplify in order to understand. This is also an important point. Human beings cannot represent the full complexity of events and form simple models. However, simplicity may have an enormous cost if we cross some delicate boundary of oversimplification. This is why I repeatedly advocate a critical and cautious approach, supported but not limited by simple models.

The Individual and the Collective

The failure to acknowledge the unique behavior of the aggregate entails fallacies of understanding and prediction. Although we may retrospectively explain the behavior of collectives, narratives can always be given post-factum. Our ability to tell stories weaving together the elements of reality and fantasy is no less than impressive. Bruner (2004) observed that human beings are gifted storytellers. However, in a complex world that West (2016) described as uncertain, unfair, and unequal, stories are no substitute for scientific representations. Commentators on China's foreign policy can provide us with narratives to frame our understanding of China, but who can predict whether and *when* the Chinese superpower will hit a tipping point of growth or turn against Taiwan?

Telling stories is a part of human nature, whatever that may be. However, telling stories and interpreting the world are no substitute for *pragmatically motivated understanding*. The phrase "the proof of the pudding is in the eating" explains this approach. It means that true value or quality can only be judged when something is put to use or tested. Pragmatically motivated understanding means that reality is the ultimate judge of our models. And when we face this judge, we will not be assessed on the aesthetic value of our stories, but on the practical consequences of our models. Therefore, a pragmatically oriented understanding refers to an approach or interpretation focused on practicality and real-world application. In this context, "pragmatic" relates to a practical, hands-on, and results-driven perspective, as opposed to one that is purely theoretical or abstract. A pragmatically oriented understanding emphasizes the importance of considering how concepts, ideas, or theories can be applied effectively in practical situations. It involves looking at the *tangible* outcomes and consequences of a particular approach or concept.

To better understand this approach, consider a recent trend where traditional martial artists are challenged and empirically "tested."[3] Those who have watched action movies as teenagers may have been impressed by the secret power of old Chinese Kung Fu masters to beat their opponents with ease, grace, and mastery, using hidden sources of "energy." When tested in reality against the brutal force of MMA fighters, these fraudulent masters are squashed like flies. Their "theory" does not stand up to the brutal reality. This pragmatically oriented approach is not limited to martial arts, of course. One should read Mandelbrot's "The (mis)behavior of markets" (Mandelbrot & Hudson, 2007) to understand how economic models did not survive the brutal realities of financial markets.

Luckily, we have gained much knowledge about the dynamics of nonlinear systems that can guide us, at least *negatively*, to a modest understanding of crowds. The main focus of the present book is not this form of understanding, though, but rather the question of how the individual *within* the collective can act non-stupidly in the face of the madding crowd. This challenge involves some deep questions concerning uncertainty and the degrees of freedom within the collective. To address this challenge, we must understand that there is an inherent difficulty in the scientific modeling of uncertainty.

Uncertainty is defined only for a large collective of particles (Lawrence, 2019). It is not defined for a single particle, and talking about the uncertainty of a single particle as decontextualized from the collective is meaningless. Adopting this perspective, the human particle, striving to understand himself within the collective, can only use the *Copernican principle*, according to which he has no unique position in existence and can reflect on himself only through the eyes of the collective. On the other hand, our understanding of ourselves as individuals who strive to have some control of our lives leaves us frustrated, not to say pessimistic and depressed, when our ignorance is conceptualized at the macro level alone. This tension is expressed through the tension between the individual and the crowd and the degrees of freedom

[3] https://en.wikipedia.org/wiki/xu_xiaodong.

and choice one has within a collective. Are we a part of the system that we observe? We may use statistical mechanics to model the system if the answer is negative. However, we may use quantum mechanics to model ourselves within the crowd if the answer is positive. And good luck to us if we do …

Dealing with our uncertainty as individuals through a concept that describes the collective seems paradoxical. Why should someone consider himself a particle among particles when he only wants to celebrate his individuality? How can someone's free will be expressed through the logic of the collective? The idea of free will is one of the traps facing us when we try to understand the individual within the collective. As explained by Rovelli (2022, p. 626), "failing to distinguish a rigid (mistaken) understanding of a concept from the actual (fluid) role that it plays within our overall conceptual structure" is a repeated fallacy.

In this book, I propose nothing contradicting our scientific understanding of uncertainty. However, I suggest that collectives have their *Achilles heels* and that individuals can study and use them to bet against the crowd. Our freedom is, therefore, expressed by playing on the outskirts of the main mass of the distribution while exploiting the variability of complex systems. This proposal reframes our understanding, reminding us that "we are all unique but never alone" (Holquist, in Bakhtin, 1990, p. xxvi). Understanding the distribution of which we are a part is the first step in understanding what it means to be different. As *variability* and *change* are inherent in all living systems, betting against the crowd means playing dynamically to change our position and exploiting "pockets" of potential freedom within and against the crowd. In this book, I use the expression "betting against the crowd" to describe the individual's attempt to adopt a contrarian approach by adopting this dynamic form of play.

Constraints and Opportunities

Betting against the crowd is a tricky business. The default is to bet *with* the crowd, as the crowd dictates our reality. Betting against the crowd pushing the stock market upward is a mistake, as it is better to ride on the wave than to bet on the exact timing of its collapse. Joining the regime is opportunistically better than fighting against it. However, in some contexts, we may want to express our individuality and gain some edge by betting against the crowd. In these contexts, we must understand the system's constraints to identify the best timing and context to take action. Constraints are limitations reducing the degrees of freedom available to the system. But constraints do not necessarily tell us where the system is heading. Negatively, and as discussed in the next section, they may mark out the system's boundary, telling us where it cannot go.

Let me use the Brazil nut paradox to illustrate the meaning of constraints. The paradox, also known as "granular convection" or "muesli effect," is observed in mixed granular materials. It describes the tendency of larger particles to rise to the top of a container filled with a mixture of particles of different sizes when the container is shaken. When you shake a jar of granola, you may notice that the Brazil nuts rise to the top, although they are larger and heavier than other components like raisins. At the same time, smaller particles settle towards the bottom. This is a counterintuitive effect because one might expect that the larger and heavier particles would sink rather than rise.

I am fascinated by this paradox, as it shows how a nontrivial pattern emerges by simply shaking a system. This emerging pattern has also fascinated physicists (Aranson & Tsimring, 2006). When you think about it, the Brazil nut paradox involves two important processes. First, the jar is shaken. No part of the action is intended to move the nuts to the top. On the contrary, the interaction between the shaking hand and the jar aims to mix the granola and increase the homogeneity of the distribution so that each bite will represent a variety of tastes. The order produced by the shaking is unintentional and even contradicts our intention to mix the components. The various particles have degrees of freedom to move in the jar. Therefore, shaking the jar should have led to a perfect mixture, maximizing the particles' entropy (i.e., variability).

This is where constraints get into the picture—size constraints. When they have been thrown up in the air and are falling back down again, the small particles sneak in between the bigger ones. Given their size, the Brazil nuts do not display the same behavior. Therefore, it is not the case that the Brazil nuts rise to the top. Rather, the smaller particles sneak to the bottom because of size constraints. The constraints present *differential restrictions* on the particles' degrees of freedom. *Order is produced when some unconstrained potential is restricted in a law-like way, given constraints formed through interactions*. The shaking hand inserts energy into the jar. The collective of particles has no intention or "essence" directing it to the formation of order. Order is generated when the energy-driven movement of the particles is constrained in a law-like way.

Considerations about constraints may be used to bet against the crowd. In Chap. 4, I explain how the identification of short time frames where hidden regularity is expressed may be important for short-term prediction. Similarly, the identification of time frames where constraints are loosened may also be important to spot when the collective starts moving away from a given trend, and to exploit this situation to bet against the crowd. In both cases, *context* and *timing* are crucial.

A contextual representation is built into living systems, from the immune system (Cohen, 2000) to human personality (Mischel, 2004) and state-of-the-art large language models (LLM) such as GPT. The point here is that humans do not respond to stimuli using general and automatic patterns. We respond with sensitivity to the context. Most people, for instance, are not pure introverts or extroverts. Some contexts may turn our gaze inside, while others may orient us toward the outside. As Mischel (2004) explains, personality is not an underlying essence hidden in our skull, but a pattern of relatively stable person–context relationships. A highly verbal individual may respond as an extrovert to contexts where he feels secure in expressing himself. However, in other contexts where he feels insecure, he may withdraw and appear as an introvert.

Contextual representation allows us to respond *adaptively*, rather than mechanically, to changes in our external and internal environment. Without contextual representation, we cannot have the variability of response which is necessary for adaptive behavior. However, variability must be constrained in a law-like manner to avoid ending up in a fully disordered state. In other words, between fully rigid and fully disordered behaviors, there is an optimal level of variability that uniquely characterizes each living system (Neuman, 2021). Adopting this perspective, we may make the following definition:

Context is the set of signals that support *constrained variability* in a law-like way.

Given our shortcomings in providing exact pointwise predictions of crowd behavior, an important idea that I would like to present in this book is that our analysis should focus on identifying informative cues or signals. These signals are contextual cues that may support our adaptive behavior. By recognizing these signs in time, an individual may use them to gain an edge, as explained in Chap. 2.

The Importance of Learned Ignorance

> In such complex businesses, you almost have no other way. You have to start with the simplest probabilities and hope to get less and less tangled later on. After all, Ariadne does not always wait for you at the entrance of the labyrinth, handing you the end of the thread. (Meiran, 1990, p. 89)

Complex behaviors do not easily lend themselves to prediction and control. In fact, this is the very thing that defines complexity, so we need to think about it differently. Years ago, I visited a leading university where the head of the institute presented a new and highly expensive predictive system sponsored by the government. I do not remember the exact details, but he tried to impress us by saying that the system predicted the second Lebanon war between Israel and the terror organization Hizballah with a 0.60 probability. I explained to him that, for the Israel Defense Forces (IDF), the working assumption is that a war is more probable than not probable, so a single pointwise probability of 0.6 would be of little help to decision-makers. After all, what does it mean to say that there is a 0.60 probability that a war is expected? That a war is more probable than getting heads when you flip a coin? What is the margin of error in this prediction? Moreover, what does it mean to the people with their boots on the ground who have to make practical decisions about things like recruitment of reserve forces?

I realized then that it is much easier to learn what to avoid than what to actually do in many real-world and complex situations, and that academic research is sometimes so blind to its shortcomings that it becomes meaningless. Saying something positive about the behavior of a complex system is not trivial, and few warriors of academia would survive if tested in the sparring ring of reality.

The idea of via *negativa* originated from theology, where it was argued that we could say nothing positive about God. Nicholas of Cusa (b. 1401) is a scholar I still remember from my undergraduate studies in philosophy. He introduced the idea of *docta ignorantia* ('learned ignorance'), emphasizing the importance of being aware of our limitations. Knowing our limits and avoiding stupidity may be the first step in understanding collectives and our own role within the collective. Knowing the system's limits means understanding the constraints operating on the system. By understanding some of these constraints, we may incrementally and modestly reduce our ignorance. The above quote from Meiran's novel "South of Antarctica" exemplifies this understanding, albeit in a different context. It describes some real-life situations where complexity dominates the situation. This complexity is portrayed as a maze from which there is no simple way to escape.

In Greek mythology, Ariadne was the daughter of King Minos of Crete. She fell in love with the hero Theseus. Theseus decided to kill the Minotaur, remembered in Picasso's paintings as a monster, half man and half beast. The monster that excited the imagination of Picasso lived in a labyrinth, so Ariadne gave Theseus a ball of yarn, which he unwound as he entered the labyrinth. After killing the monster, the hero used Ariadne's thread to find his way out of the maze and back to his lover. Even if we are Greek heroes, we must seek assistance to get out of a maze.

Similarly, when studying complex systems, such as crowds, we are forced to start with simple assumptions and should "hope to get less and less tangled later on." The notion of *docta ignorantia* is important primarily for its didactic value as a guiding principle. Suppose we do not really understand the "essence" of the collective. In that case, we can only approach it negatively, accepting that we cannot provide a precise prediction of how far the stock market would fall under the rush and roar of the madding crowd. Is there a way in which we can say something positive about what to do? This is the challenge that will be discussed throughout this book.

The Dancing Crowd

> I played and replayed the scenes … trying to find some order, pattern. I found none. (Didion, 1971, p. 12)

When I dance with my two-year-old granddaughters, we dance in a circle or a straight line, pretending to be a train. We sing and move symmetrically, either through rotation or translation. These dances express simple and predictable order, which is adaptive to my granddaughters' developmental level and my limited dancing skills. Dance patterns range from the highly ordered to the highly disordered, where the madding crowd behaves like gas particles (Silverberg et al., 2013). My dance with my granddaughters is fully ordered. Rave dance and moshing are total chaos. Modern dance, as epitomized in the choreography of Alonzo King, is non-simple, complex, and unpredictable. So, which form of dance would be best for a metaphorical description of crowd behavior?

The kinds of dance described here represent three prototypical forms of behavior: chaotic, ordered, and non-simple. In a deep sense, both chaotic/random behavior and ordered behavior are simple to model. In the first case, we can model the dancers' behavior as an example of Brownian motion, where the particles are pushed around randomly by gas particles, joyfully boosted as the temperature gets higher. In the second case, we can model the behavior of the particles by analyzing the identity-preserving transformation of their structure. If you have watched the military marches of fascist regimes like North Korea, you can easily identify the structure and dynamics of the soldiers' movements. Dictators are fond of order and predictability and have no patience with the expression of variability. Simplicity and complexity are two extremes that are easy to identify. We all recognize the difference between the ordered march of soldiers and the disordered movements of free dancers at a rock festival. Interestingly, at some music festivals organized by neo-Nazis, the crowds dance in a totally disordered way. Extreme order and total disorder seem to have something in common: they do not like life as it is. In contrast with these two extreme forms, deciphering the thinking behind Alonzo King's "Meyer" is challenging. Collectives may present all three forms of behavior. However, for the individual participating in the dance of the collective, the important question is not what the pattern is, but *how variability in the pattern* may be used as an opportunity for betting against the crowd.

Dance is a nice metaphor, but human behavior differs from dance. Like the Brazil nut paradox, it may display non-trivial behavior when subject to changing constraints. Dancing with a crowd is much more challenging than dancing on stage. For the individual dancer in King's Meir, there is a well-defined choreography that is opaque to the audience. She cannot simply make her own moves. For the observer, the complexity in work is not the same as the complexity for the performing dancers. Whether choreography underlies our collective behavior is an interesting question. Complex social systems, like crowds, seem to express different forms of order.

Scientific Thinking in the Absence of Truth

> Fortunately, in the absence of truth, facts take its place, and it is difficult to contend with them. (Meiran, 1990, p. 74)

A naïve conception of science presents it as lifting the veil to uncover nature's true face. This may be true of physics and biology. However, "truth" is an idea that is much more difficult to understand than facts. Facts are generally considered more concrete and less subject to interpretation or personal belief. They are often based on empirical evidence and can be supported by data or observations. In contrast, truth can sometimes be relative or influenced by individual perspectives. Facts are typically objective and can be objectively demonstrated or proven. For example, the statement "The Earth orbits the Sun" is supported by scientific evidence and observations. However, a statement like "chocolate is the best ice cream flavor" may be true for someone who particularly enjoys chocolate. But it is not a fact, because it is based on personal preference and not universally verifiable.

The difficulty in gaining access to "truth" is not a barrier for knowledge seekers. Maybe this is why the legendary John von Neumann said that science does not try to explain or interpret, but places *models* above all else. This is especially relevant when modeling collectives and the individuals within them. The "truth" underlying the behavior of a collective may be hard to decipher, but there are facts, measurable and proven behaviors, that can be used to model the collective. We use facts as the building blocks for a model which is always a simple representation of the complex behavior we seek to understand.

Jorge Luis Borges wrote a short story titled "On Exactitude in Science" (Borges, 1999). It's a brief but thought-provoking tale. In Borges' story, he describes an empire where the art of cartography has reached an extreme level of perfection. The mapmakers of this empire have created a map so detailed and accurate that it is an exact 1:1 scale replica of the entire territory it represents. In other words, the map covers every square inch of the land it represents, leaving no room for gaps or discrepancies. This colossal map is so large and unwieldy that it cannot be used for navigation. In fact, it is so immense that it lies in tatters across the empire, with people using various portions of it for different purposes. Some people live on sections of the map, others study it as a scholarly pursuit, and some even use it as a flag.

The tale is often interpreted as a parable about the nature of knowledge and representation. It raises questions about the relationship between reality and our attempts to represent it, suggesting that perfect representation is impractical and perhaps even impossible. The map, intended to represent reality, merely becomes a burden and a symbol of the hubris of those seeking to create a perfect world model.

Science is about models, explained von Neumann, and a model is an abstract, simple, and clear representation aiming to improve prediction and control over the *best practice*. The last point is highly important for a pragmatic approach. As argued by Zilliak (2019, p. 283):

> Whatever the purpose of the experiment, best practice research compares a novel treatment or variable with best practice and/or prevailing wisdom, not with an assumed-to-be-true null hypothesis or blank placebo.

To prove the relevance of our model, we must compare it with the "best practice," whatever that may be. When betting against the crowd, we may consult the "best practice" we know of. Being able to model the crowd's behavior provides the individual with an edge over the best practice, but not with access to the "truth."

The Rebel's Perspective

I must emphasize that this book is not a philosophical treatise. It attempts to understand some *foundational* scientific aspects of crowds and describe how the *empowered* individual may find his way *within* such a collective by understanding the underlying dynamics. The book thus combines a scientific attempt to model collectives in a way that may be useful for the individual to find his way within, and largely against, the crowd. In this sense, it is a scientific contemplation for the rebel who strives to gain individuality while confronting the madding crowd. And it is no coincidence that the book is being written at a time when Israeli democracy is facing its most difficult challenge. The interaction of dark political forces has formed a soliton, moving forward against the only democracy in the Middle East and the history of the Jewish people. This wave, which has surprised most Israelis, including myself, is nothing less than the perfect context for the current book.

Several words should be said about the general approach used in the book and the audience for which it is intended. The book is scientific. However, it is written as a self-contained book for the educated reader. It aims at a wide audience interested in understanding the dynamics of crowds. Moreover, it specifically focuses on the way individuality may be empowered within such collectives by adopting a scientific and reflective approach. No recipes are given, just guidelines which are supported by a certain amount of scientific evidence.

Monty Python's "Life of Brian"[4] includes a hilarious scene in which a poor individual is to be stoned to death. The bloodthirsty crowd includes women wearing fake beards because women are forbidden to watch such an event. The humor in this scene does not hide the fact that the crowd, as observed by Le Bon (Le Bon, 1895, p. 17), "is guided almost exclusively by unconscious motives" (some of which may be dark and dangerous). The empowerment and emancipation of the individual is a process that must vanquish the barrier of the unconscious. I invite the reader on a journey of contemplation where she will meet different crowds, from Hungarian political parties to groups of gamblers and stock market traders. These are cases where a lesson could be learned and hopefully empower the individual in the face of the madding crowd.

[4] https://en.wikipedia.org/wiki/Monty_Python%27s_Life_of_Brian.

The Structure of the Book

The book draws on some of my previous more technical papers (Neuman & Cohen, 2022, 2023; Neuman et al., 2021), but the themes presented are original and mostly appear in the first, more general part of the book. In fact, there are two parts. Part I is introductory and aims to present general ideas. Part II is more specific and technical and aims to analyze different examples and aspects of crowd behavior and to deepen our understanding of how the individual may find her place within and in opposition to the crowd. The reader may find these two parts quite different in their orientation. However, he may benefit from reading the more technical aspects of the book as these are the places where ideas are made explicit. Speaking about technicalities, I must conclude by apologizing for any technical errors. Throughout my academic career, I have noticed that regardless of careful and repeated reading of the manuscript, it is almost inevitable that some error will slip into the text. Given the thesis presented in the book, such surprises are inevitable and almost a normal part of any natural system that is not artificially designed for mass production.

References

Aranson, I. S., & Tsimring, L. S. (2006). Patterns and collective behavior in granular media: Theoretical concepts. *Reviews of Modern Physics, 78*(2), 641.

Bakhtin, M. M. (1990). *Art and answerability: Early philosophical essays*. University of Texas Press.

Bakhtin, M. M. (1984). *Rabelais and his world* (Vol. 341). Indiana University Press.

Bateson, G. (2000). *Steps to an ecology of mind*. University of Chicago Press.

Borges, J. L. (1999). On Exactitude in Science. Borges JL Collected Fictions (trans. Andrew Hurley).

Bruner, J. (2004). Life as narrative. *Social Research: An International Quarterly, 71*(3), 691–710.

Cohen, I. (2000). *Tending Adam's garden: Evolving the cognitive immune self*. Elsevier.

Didion, J. (1971). *Play it as it lays*. Bantam Books.

Kosiński, J. (1965). The Painted Bird. Houghton Mifflin.

Lawrence, A. (2019). Probability in physics. In *Undergraduate lecture notes in physics*. Springer, https://doi.org/10.1007/978-3-030-04544-9.

Le Bon, G. (1895). *The crowd: A study of the popular mind*. T. Fisher Unwin.

Mandelbrot, B., & Hudson, R. L. (2007). *The misbehavior of markets: A fractal view of financial turbulence*. Basic books.

Meiran, R. (1990). *South of Antarctica*. Keter. (in Hebrew).

Mischel, W. (2004). Toward an integrative science of the person. *Annual Review of Psychology, 55*, 1–22.

Neuman, Y. (2021). *How small social systems work: from soccer teams to jazz trios and families*. Springer Nature.

Neuman, Y., & Cohen, Y. (2023). Unveiling herd behavior in financial markets. *Journal of Statistical Mechanics: Theory and Experiment, 2023*(8), 083407.

Neuman, Y., Cohen, Y., & Tamir, B. (2021). Short-term prediction through ordinal patterns. *Royal Society Open Science, 8*(1), 201011.

Neuman, Y., & Cohen, Y. (2022). A permutation-based heuristic for buy low, Sell High. arXiv preprint arXiv:2207.01245.

Roser, M., & Ritchie, H. (2018) Optimism and pessimism. Published online at OurWorldInData.org. Retrieved from: https://ourworldindata.org/optimism-and-pessimism [Online Resource] https://ourworldindata.org/optimism-and-pessimism

Rovelli, C. (2019). *The order of time*. Penguin.

Rovelli, C. (2022). The old fisherman's mistake. *Metaphilosophy, 53*(5), 623–631.

Silverberg, J. L., Bierbaum, M., Sethna, J. P., & Cohen, I. (2013). Collective motion of humans in mosh and circle pits at heavy metal concerts. *Physical Review Letters, 110*(22), 228701.

West, B. J. (2016). *Simplifying complexity: Life is uncertain, unfair and unequal*. Bentham Science Publishers.

Ziliak, S. T. (2019). How large are your G-values? Try Gosset's Guinnessometrics when a little "p" is not enough. *The American Statistician, 73*(sup1), 281–290.

2

Signs of Collective Dynamics: Insights from the Stock Market Collapse

Introduction

From the early'20s, the New York stock market and its main performance index, the Dow Jones, showed unprecedented growth. "From 1920 to 1929, stocks more than quadrupled in value,"[1] meaning that investors increased their wealth fourfold. For example, if an investor bought a stock for $100, it would be worth $400 nine years later. Selling his stock at this point would leave the investor with a profit of $300. Such an increase in wealth would have been considered unthinkable by most Americans living at that time. For a better understanding of the excitement this inspired, take a look at Fig. 2.1 showing the increase of the Dow Jones index from 1 January 1920 to 1 September 1929. The X-axis shows the timeline, and the Y-axis the Dow Jones index.

[1] https://www.pbs.org/fmc/timeline/estockmktcrash.htm#:~:text=From%201920%20to%201929%20stocks,an%20even%20more%20precipitous%20cliff.

Y. Neuman, *Betting Against the Crowd*, https://doi.org/10.1007/978-3-031-52019-8_2

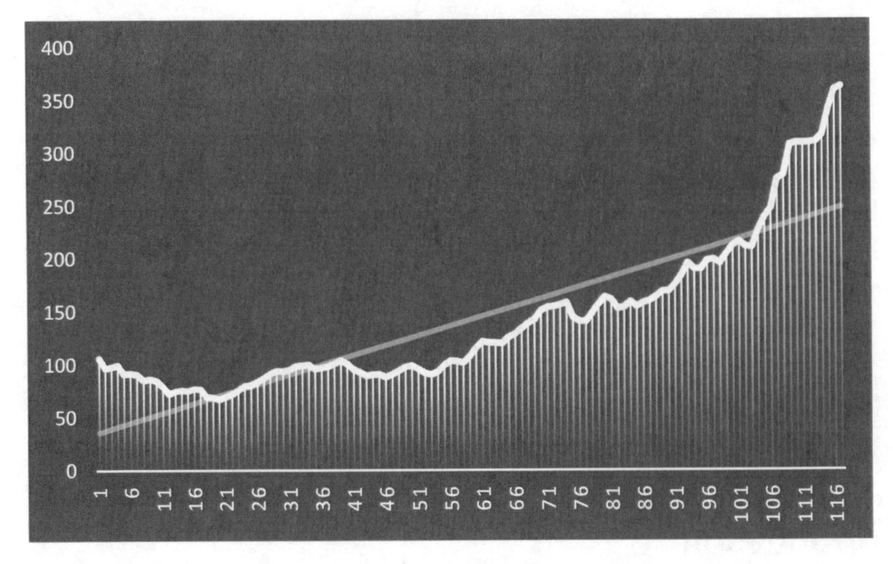

Fig. 2.1 Change in the Dow Jones index during the 1920s. *Source* Author

You can see that, beyond a certain point, the increase follows a general pattern described as *exponential growth*. This kind of growth is represented by the following equation:

$$y = a(1 + r)^t$$

where y is the future amount or value, a is the value we begin with, r is the growth rate in decimal form, and t is time. For example, if you have a creature that duplicates every day, in which case $r = 1$, and you start with a single creature, then on the second day you will have 2 creatures, on the third day 4 creatures, and so on.

The straight line in the above figure represents linear growth, where a constant value is added at each step. This shows us how the actual growth of the index deviates from linear growth. The magic of exponential growth is that it involves the proportional growth of a value as a function of discrete time steps. The values you observe in the next step increase with respect to the previous value. For example, bacterial growth is exponential, which is why a restaurant kitchen with low hygiene standards is so dangerous. Even an extremely small number of bacteria in food can rapidly grow to huge numbers, causing a risk of food poisoning.

Here is another example. If we start trading stock with an initial value of $1 and anticipate exponential growth with $r = 1$ (100%), then in the following steps, our stock should be worth $2, $4, and so on. Exponential growth is impressive, precisely because we observe a proportional rather than linear increase in value. The value increases in proportion to its value at the previous time step.

The psychological aspect of exponential growth can be understood if we take into account the fact that our mind has evolved to work logarithmically (Varshney & Sun, 2013) rather than linearly or exponentially. Let me explain this point. The logarithm of a number using a base "b" is the *exponent* to which b should be raised to produce the number. For instance, if we have a base of 10, we should raise it to the power of 2 to get the number 100. Therefore, $\log_{10}$ of 100 is 2. What happens if we calculate 10 to the power of 3? When we raised 10 to the power of 2, we got 10 * 10, which is 100. When we compute 10^3 we get $10 * 10 * 10 = 1000$. Increasing the exponent by just one point from 2 to 3 did not create a corresponding increase in the outcome. We increased the number from 2 to 3, but the outcome was 10 times higher than the previous outcome.

Transforming numbers using the logarithmic transformation means we map our numbers to a logarithmic *scale*. The logarithmic transformation is non-linear. On a line with a linear scale, every unit of distance is generated by adding the *same* amount. The distance between 10 and 100 is generated by adding 90 units, and the same for the distance between 9910 and 10,000. In contrast, using a logarithmic scale, every unit of length is generated by *multiplying* the previous value by the same value, which means that the numbers are not equally spaced. This is important because logarithmic transformations have several benefits, such as compressing larger values. Logarithmic scales can also accentuate small differences in values that might be obscured on a linear scale. By compressing larger values, logarithmic scales allow for better resolution of smaller values, making it easier to distinguish subtle variations or deviations from a baseline.

At this point, we may better understand why the human mind is logarithmic. As natural beings, we are not interested in mathematical quantities, only in meaningful differences. Think, for example, about buying a good whisky. Unless you are an expert with highly sophisticated taste or a *Nouveau riche* trying to impress his social milieu with a highly expensive Scotch, your main interest will be in differentiating between good and bad whisky. This is the most basic difference for all living systems, a difference between good and bad, a difference that makes a difference. When you visit the liquor store, you may be offered a variety of Scotch whiskies, ranging from $10 to $10,000 a bottle. Is the difference in quality between the whiskies costing $10 and $100 the same as the difference in quality between the whiskies costing $9910 and $10,000? While the mathematical difference of 90 units is the same, the *meaning* of this difference differs for the two extremes on the price scale. There is a more significant difference in taste between the cheapest and the $100 whisky. This "difference of differences" can be represented if we transform the prices using a base-10 log:

$$\text{Log}_{10}10 = 1$$
$$\text{Log}_{10}100 = 2$$

$$\text{Log}_{10}9910 = 3.996$$
$$\text{Log}_{10}10,000 = 4$$

As you can see, using a logarithmic scale, the $100 Scotch is represented by twice the value of the $10 whisky, while there is a negligible difference between the $9910 and $10,000 whiskies. Logarithmic transformation allows us to expose a difference that makes a difference and ignore one that does not.

In evolutionary adaptive terms, we are far more concerned with significant thresholds and meaningful differences than purely mathematical differences. For example, suppose you are the legendary Indiana Jones visiting an exotic Island and you are suddenly attacked by a bunch of bloodthirsty cannibals. In that case, you are more concerned by the difference between a non-attack (no cannibals are chasing me) and an attack (three cannibals are chasing me) than by the absolute number of cannibals *beyond* a certain threshold. Whether 200 or 300 cannibals chase Indiana Jones makes no difference, as in both cases, his escape response is the same. A difference that makes a difference is a difference that entails a certain response. In the realm of living organisms, "*meaning*" is the basic unit of measurement, and "meaning" means a difference that makes a difference. If, and for all practical purposes, there is no difference between 200 and 300 cannibals chasing Indiana Jones, then the absolute difference of 100 cannibals is meaningless.

Like the minds of all living creatures, the human mind is not interested in differences as they actually are, but only in *meaningful* differences that entail a certain response accompanied by a price tag. Going back to Indiana Jones, whether he is or is not being chased by cannibals is a difference that makes a difference (Bateson, 2000). Programmed by this logic, it is no surprise that we find it difficult to understand the meaning of exponential or super-exponential growth and its consequences.[2] While our logarithmic mind compresses large differences, exponential growth amplifies them, in a way, our simple minds find it difficult to understand. This is a potential source of

[2] In cases of super-exponential growth, the growth rate increases as the system becomes larger.

danger. Our mind has learned to compress differences at the high end of the scale. However, this is exactly where exponential growth may reach its limits and collapse.

In this sense, we are like the frog in the famous "*Boiling Frog Parable.*" According to the parable, if you put a frog into a pot of boiling water, it will immediately jump out to save itself. However, if you put a frog into a pot of water at room temperature and then slowly heat it, the frog will not perceive the danger and will stay in the pot. As the water temperature gradually increases, the frog will adapt to the changing environment until it is too late and can no longer escape.

Exponential growth does not involve small incremental differences. However, an important lesson we have learned so far is that certain *representations* of the world entail different views of the world, and different views entail different forms of behavior. The poor frog loses its life because its representation of the body heat is insensitive to incremental changes that can be accommodated, and it ends in disaster. To get noticed, a temperature change must cross a threshold known in psychophysics as the *just noticeable difference* (JND). If the change is smaller than the JND, it goes undetected.

Despite their long and impressive evolutionary history, frogs have not evolved to adapt to life in the cooking pot. When we deal with the behavior of crowds, we may assume that, despite the impressive flexibility of the human mind, there are contexts where it exemplifies the boiling frog syndrome. Habituation to incremental and undetected changes may lead to disaster. If you are a reflective frog, you may ask yourself a simple question: Is this a situation I am naturally adapted to? If the answer is negative, then steps must be taken immediately. Ideally, the frog should carry out a quick calculation, extrapolating to the temperature of boiling water. It would then understand that it would be unable to withstand such a temperature.

Body temperature is of specific interest for another reason. It turns out that we do not have thermostats in our brains. Rather, our brain senses a dangerous increase in body temperature through a network of sensors and signs indicating something bad is happening. These signs are not a direct measurement of the temperature; observing them, we do not somehow "see" a number indicating our body temperature. They are signals, registered biological activity, interpreted by the brain as signs of danger. When a crowd of people start investing as they observe exponential growth, they may be blind to its meaning and, even more importantly, to signs of danger. In this context, betting against the crowd means being a reflective frog, seeking a way out of the water while the rest of the crowd of optimistic and cheerful frogs just go on enjoying the hot bath.

Let us return to exponential growth. Our mind is logarithmic, emphasizing small differences and compressing big ones. As such, we may be impressed by exponential growth but, at the same time, insensitive to its meaning and consequences. This is probably one reason the stock market machinery looked like magic for ordinary Americans striving to make a living. It seemed that there was no need to sweat a single drop in hard labor to make a living. You could just buy a share that would miraculously increase in value. Like alchemy, straw would turn into gold, exponentially increasing its value beyond anything our simple minds could understand.

Seeing the easy money that could be won on the stock market, ordinary Americans joined the game. The rush for stocks was so intense that "an estimated one person in a hundred in the population was playing stocks" (Whalen, 1993, p. 101). This is an amazing figure, given the conditions people were living in at the time. In its report "100 Years of U.S. Consumer Spending", the U.S. Department of Labor shows[3] that, at the beginning of the twentieth century, most of a family's income would be spent on necessities: food (43%), housing (23%), and apparel (14%).[4] Therefore, the rush to the stock market involved too many households with almost no *surplus*, precisely because most of their income was spent on necessities. But, unfortunately, too many families invested in the stock market.

All living systems involve some *redundancy* as a safeguard against failure. When there is no redundancy, the system may be in danger when it faces risk. Americans investing with no such redundancy, no surplus financial resources, were clearly at risk. In retrospect, this situation seems trivially dangerous. There were too many individuals playing a risky game with no surplus that could cover them for this risk. Still, hindsight wisdom is always just hindsight wisdom rather than foresight wisdom, and as such, is of little use to those involved. We should remember that the years of the growing stock market, described as the "*Roaring Twenties*" or the "*Jazz Age*," were years of economic growth, optimism, changing social order, and new opportunities. Optimism was the *Zeitgeist*, and Zeitgeist is always stronger than any form of rationality. We will discuss the optimism of crowds in a later chapter. The subject will be crowds of football fans. But for the moment, let us reflect on ways that simple emotions and complex emotion-based representations of the future, such as optimism, might distract us from any form of "rational" thinking.

---

[3] https://www.bls.gov/opub/100-years-of-U-S-consumer-spending.pdf.

[4] https://www.theatlantic.com/business/archive/2012/04/how-america-spends-money-100-years-in-the-life-of-the-family-budget/255475/.

At this point, we should note that the concept of rationality is less clear than we tend to believe. In fact, the simple idea of non-stupidity may better fit our discussion. The idea is that ultimate stupidity (Cipolla, 2021) involves damage to you and others. If you are a taxi driver from New York who has spent all his wealth on the stock market, you and your poor family are destined to encounter poverty on the day of the collapse. Rationality has been elaborated by wise guys, mainly philosophers, psychologists, and economists. But rationality has one very basic meaning: do not be stupid.

The wise Charlie Munger once reflected on his success by saying, "It is remarkable how much long-term advantage people like us have gotten by trying to be consistently *not stupid*, instead of trying to be very intelligent." (Beveling, 2007, p. 328, my emphasis). Trying to be "not stupid" is an ideal I warmly embrace, and I am repeatedly amazed at how difficult and challenging it is to be "not stupid." Personally, I have failed to meet this challenge on more than one occasion ...

It is important to emphasize that being "not stupid" is not the same as being intelligent. Think, for example, about nutrition, which is a major source of interest for the neurotic Westerner. There is much stronger evidence for the "do not" of healthy nutrition than the "do" of healthy nutrition. Intelligent nutrition requires complexity and sophistication beyond the scope of the common individual. Matching the exact quantity of proteins, fats, carbohydrates, minerals, and vitamins against the individual's specific genetic, gender, and age profiles is something that goes beyond our understanding. However, advising the elderly to avoid highly processed food in unreasonable quantities that do not correspond to their energy expenditure is good enough advice for those seeking non-stupidity, and this is a feasible ideal for most of us. Non-stupidity should be our default strategy, while intelligence and "rationality" can come later if correctly interpreted and used.

Back to history and the unhappy end of the roaring stock market. *It simply collapsed*. The collapse started in September 1929 and was steep and painful, as can be seen in Fig. 2.2.

"The Dow then embarked on another, much longer, steady slide from April 1930 to July 8, 1932, when it closed at 41.22, its lowest level of the twentieth century, concluding an 89.2% loss for the index in less than three years".[5] In the context of the Roaring Twenties and the Zeitgeist of optimism,

[5] https://en.wikipedia.org/wiki/Wall_Street_Crash_of_1929#cite_note-25.

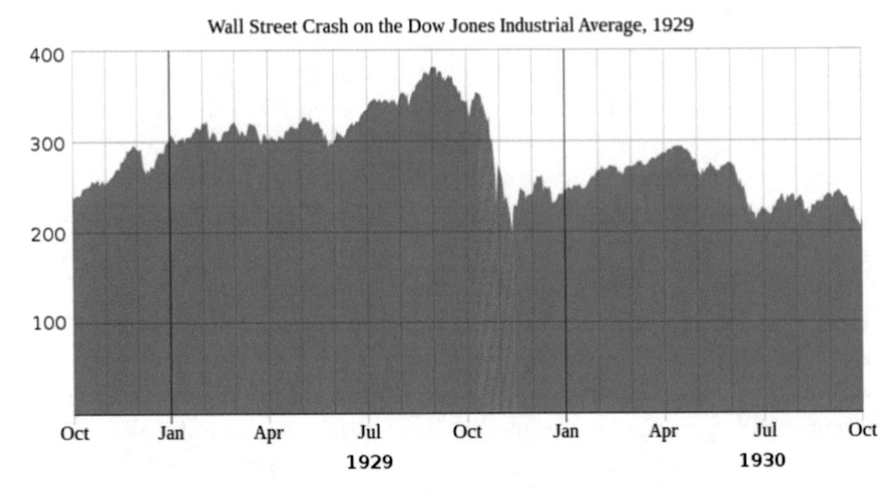

Fig. 2.2 The collapse of the stock market. *Source* Wikipedia

the stock market collapse traumatized American society. The Great Depression that followed is registered in our collective memory through pictures[6] of poverty, unemployment, and despair.

Surprisingly, one notable individual escaped the collapse of the stock market: Joseph Kennedy, the famous businessman, investor, and father of President John Kennedy. A famous story is that Kennedy decided to bet against the crowd (Whalen, 1993, p. 100) after getting investment advice from a shoeshine boy. He "decided that a market anyone could play, and a shoeshine boy could predict was no market for him" (ibid, p. 105). Another prominent figure who left the market for surprisingly similar reasons was Bernard Baruch, who said:

> Taxi drivers told you what to buy. The shoeshine boy could give you a summary of the day's financial news as he worked with rag and polish. An old beggar who regularly patrolled the street in front of my office now gave me tips and, I suppose, spent the money I and others gave him in the market. My cook had a brokerage account and followed the ticker closely. Her paper profits were quickly blown away in the gale of 1929. (Cited in Rothchild, 1996)

[6] https://www.smithsonianmag.com/smart-news/meet-photographers-charged-documenting-depression-era-america-farm-security-administration-180964123/#:~:text=Dorothea%20Lange&text=Her%20%E2%80%9CMigrant%20Mother%E2%80%9D%20photographs%20shot,photographs%20of%20the%20Great%20Depression.

What the two gentlemen saw as a bad omen deserves careful attention, far beyond the amusement of the shoeshine boy anecdote. Let us begin by adopting a skeptical perspective, given that "The Cause" of the crash is still debated. Suppose, then, that we cannot determine the cause of the crash. One may even doubt whether there is any point in looking for "The Cause" of the collapse. Sornette (2009), for instance, argued that we should look at the system's stability rather than the causes of the collapse. He argued that numerous causes can lead to collapse when a system is unstable. Think, for example, about a very old person who falls while taking a shower, breaks his hip, is hospitalized with a hip fracture, and passes away shortly after. When asked to identify the cause of death, it is clear that there are several, or rather a chain of causality that is far more complex than any simple explanation. Is it the old man's instability that led to his death? The lack of social support to help him in his daily life and movements? The loss of bone density (osteoporosis)? Or a weak immune system incapable of handling potentially dangerous fungi found in the hospital? In this context, seeking a single ultimate cause seems meaningless. The old man was an unstable system prone to collapse, and the specific trigger of the collapse was secondary to the system's overall instability.

There is an important lesson that we may learn from Sornette's argument. The world is complex, and asking simple causal questions such as why the Roman Empire or the stock market collapsed may be interesting, but it is less useful for our daily conduct and behavior. The idea of *profound simplicity* urges us to understand that looking for simple causes may not be the best strategy in certain complex contexts. Understanding the underlying *dynamics* and identifying bad omens that signal those underlying dynamics is more than enough to choose the appropriate behavior, but this simplicity may sometimes be overshadowed by senseless intellectualization.

This is a really important point for understanding crowds. The dynamics of the crowd may be complex, but there may be underlying and general patterns (e.g., exponential growth) that are informative enough to help us make non-stupid choices when taken together with certain simple signs. The idea of profound simplicity is a guiding principle in this book. Its basic assumption is that the world is too complex to fully grasp. Simplicity should, therefore, be a guiding principle. However, underneath this simplicity there is nevertheless a form of profound understanding. While it sounds paradoxical, the guiding principle of profound simplicity may be extremely constructive, as I hope to illustrate throughout this book.

What do we learn from the shoeshine boy anecdote? What can we learn about crowd dynamics and the empowered individual striving to survive within the madding crowd and its gold rush? Bernard Baruch mentioned four professions involved in the stock market: taxi driver, shoeshine boy, cook, and beggar. In these cases, we are talking about working class or even "lower" class individuals. So, is there a problem with working-class individuals or beggars enjoying the stock market boom? To understand the problem, we must first understand the stock market.

As its name suggests, the stock market is a place where certain things are bought and sold. In contrast with the old marketplace where materials and goods are traded, the stock market is a place where a company offers a portion of its ownership to investors in the form of shares. For example, General Electric was one of the 30 large companies listed on the Dow Jones index in 1920. Such a company could then recruit money from the public to continue its growth. The value of a stock increases or decreases as a function of demand. Of course, it's a good thing to be a shareholder in a successful company, that is, a company that has used its money wisely for growth, as reflected in its performance measures, since this will in turn attract more investors and increase the company's stock value.

For instance, let us assume that General Electric is expected to get a license for building and operating a new power station. This information entails an increase in profits, so the company recruits money from the public to build the new power station. Investors then benefit from the increasing value of their shares, representing the market's optimism regarding the company's future performance.

However, the idea of value is tricky, and as we learned from the logarithmic scale, our measurements do not trivially correspond with a value metric. Let me explain. Let us assume that you bought stock. Over one year, it doubled its price. You can sell the stock and earn a tidy sum, but inflation may eat your profit unless you use it for another value-based activity (e.g., buying real estate). Money (and value) has meaning only in *exchange* (Goux, 1990), so *value* is a symbolic-dynamic phenomenon that exists only in human *interactions*. It is not an objective representation of reality. It is formed through exchange and dynamics, and these are prone to all the fallacies of the crowd. As long as you keep holding the stock, it may increase or decrease in value. However, value exists only in exchange, and exchange, as a dynamic process, is prone to ups and downs as a function of public whims that may have nothing to do with the company's intrinsic value or its real measured performance.

However, not everything is psychology. The process of value growth, like any other natural growth process, may be limited from below by the scarcity of resources. After all, if you run out of buyers, those looking for a profit through increased demand will eventually lose. In a deep sense, unsustainability and limited resources play a role in the stock market, as they do in *Ponzi or pyramid schemes. I* am not saying that the stock market is a scheme, but that we may identify bad omens in the crowd's unsustainable "games" when they are played with limited resources. To understand this point, let us understand a pyramid scheme and its underlying mathematics.

Ponzi and the Pyramids

A pyramid scheme starts with a promoter, such as Madoff or a group of con artists, that recruits individuals by promising them some financial benefits. Each level of participants that joins the scheme is asked to invest its resources in the game and recruit more people. Each new level of participants "feeds" the higher level through some reward paid to the recruiters. As such, value flows *only* from the lower levels of the pyramid toward the top, in one direction. As Bateson explained (Harries-Jones, 1995), social systems are formed through a *recursive hierarchy* where information flows in both directions between different levels of the system. Unfortunately, too many people are ignorant of this very important idea. The structure of a pyramid scam violates this natural logic, which is why it is so dangerous.

Previously, we saw that evolved natural systems include redundancy as a safeguard against the turbulence of reality. A system with no redundancy is a dangerous system. A crowd operating with no redundancies on its side is in danger. Here, we learn another lesson. Information and value flow both ways in natural living systems: up and down. A system where the flow is one-directional violates the basic logic of all living systems.

As long as value flows, participants in the pyramid game may benefit. A pyramid may benefit the con artists, but it is destined to collapse because it relies on a constant influx of new participants to generate returns for existing participants. The continuous influx of new "suckers" is necessary for the generation of profit. Eventually, the process hits the bottom when the reservoir of suckers is exhausted, leading to collapse. Leaving the sinking ship in good time may be the ultimate strategy for keeping the profit in a Ponzi scheme. For example, if you joined the pyramid early and got out in time, you may enjoy a profit, unless of course law enforcement and tax agencies get their hands on you. Still, since the madding crowd playing the game is motivated by greed, such an escape plan is rarely actually applied, and most participants in a pyramid experience a significant financial loss. So, do not get involved in such activities unless you are stupid or a con artist.

The point we expect to hit the bottom of a pyramid is important if we want to work out the best timing for leaving the sinking ship. In this context, mathematics comes to our assistance, no less than a good understanding of human psychology.

Analyzing the mathematics underlying the pyramid scheme, Corden (2019) points to the exponential growth of the pyramid. If each member recruits M members below them, then each level will contain M^{N-1} members, where N is the number of the level. For example, if each recruiter has two followers, level two of the pyramid contains two members, and the overall number of participants is three. Knowing the growth rate of the pyramid and how many people each participant recruits, we can calculate the number of levels to hit the bottom of the pyramid, given the maximal size of the population that can play the game. Then it turns out that the number of the first level that will tip the total number of members of the scheme over a population of size p is

$$\frac{\ln(p+1)}{\ln(M)}$$

where "ln" denotes the logarithm with base *e*.

Let us assume that we can consider levels in terms of years. As we join the stock market, we make a rough guess that the M people recruited to the "pyramid" each year can be approximated by considering its exponential growth. As there is no sense in recruiting fewer than two people at each level, our minimal M is 2. From 1 January 1920 to 1 August 1929, the Dow Jones Index increased almost fourfold (i.e., from 105.9 to 359.15). We can therefore use $M = 4$. What is the ultimate size of population that could have participated in the stock market?

The US population in 1920 was 106,461,000 people, and each household contained on average four people, so the men in the families would have composed a population of approximately 26,615,250 people with the potential to take part in the game. This is an overestimate as not all males actually invested in the stock market. However, we use this number for illustration only. By the above equation, we get

$$\frac{\ln(26615250)}{\ln(4)} = 12$$

Assuming the pyramid began to form in 1920 and included all the relevant population of players, it can be estimated that it should reach its bottom and experience a collapse in 1932, twelve years later.

This model is built on oversimplified assumptions. After all, and in contrast with Ponzi schemes, different Americans invested different sums of money in the stock market, and the share of the rich population was much higher than the investment of the cook, the taxi driver, and the beggar. Moreover, the stock market did not necessarily have to collapse! Unlimited exponential growth is actually bounded by the scarcity of resources. Although growth may start out exponential, at a point known as the *inflection point*, where the growth rate gets closer to 1, it may change into a *logistic curve*, where a balance is reached. This is not what happened with the stock market. Therefore, even under my oversimplified assumptions, this is a nice approximation, giving a good indication of the year the stock market collapsed. To recall, "The Dow then embarked on another, much longer, steady slide from April 1930 to July 8, *1932* (emphasis mine), when it closed at 41.22, its lowest level of the twentieth century, concluding an 89.2% loss for the index in less than three years".[7] In some cases, guessing can do quite a nice job, at least in giving us rough approximations.

Again, the stock market is not a pyramid scheme. Still, the stock market and the pyramid scheme share a common denominator that is well understood today: exponential growth is unsustainable under limited resources. Whether talking about the global economy, illusory schemes repeatedly used by con artists, or the stock market, exponential growth is unsustainable under real-world constraints where available resources do not themselves support exponential or super-exponential growth (West, 2018). Moreover, a lack of redundancy and a one-directional flow of value may be signs that the system, in our case the crowd, is not resilient. Like the old man, it may collapse, given the right trigger.

The crowd motivated by greed is pushing the market upward, but here comes the math, and it tells us that this carnival of greed must end either peacefully or painfully under the condition of limited resources. The empowered individual, who understands crowd dynamics, may still enjoy the company of the madding crowd as long as she knows when to abandon ship, just before hitting the iceberg. *Being non-stupid means being sensitive to timing*

[7] https://en.wikipedia.org/wiki/Wall_Street_Crash_of_1929#cite_note-25.

and context. Moreover, being non-stupid means being sensitive to the structure of the system and its built-in bugs. A system lacking redundancy is like a mad gambler, something we shall discuss and illustrate in one of the following chapters.

How to Recognize Bad Omens

At this point, we may speculate about the bad omen identified by Kennedy and Baruch. Both might have intuitively understood the idea of exponential growth because information propagated freely, even between the upper and lower classes (i.e., businessmen and beggars), and this spread of information resulted in *herd behavior* and the market's rapid growth. The "shock" they experienced from the interaction with shoeshine boys could have come from the way information spread like wildfire. Information spreading exponentially like a rumor is too fast to control and manipulate, and for a trickster like Kennedy, a lack of control and manipulation was probably a worrying sign.

Betting against the crowd may sometimes mean distancing yourself from the crowd because you understand that you'll get no edge by being a part of it. Moreover, Kennedy and Baruch may have intuitively understood the idea of unsustainable growth and suspected that the newcomers to the game signaled the approaching bottom level of the pyramid and the tipping point of the stock market. The shoeshine boys would not have been able to recruit more investors to push the market upward. In addition, Kennedy and Baruch were sensitive and vulnerable to market turbulence. With no surplus of resources, any blow could have knocked them out, like the old man I mentioned before.

Their intuition was impressive. With no massive datasets, Excel sheets, or sophisticated quantitative models, they identified a single sign indicative of worrying underlying dynamics. Over-sophistication could have led them astray. Their critical advisors could have asked them: Beyond your anecdotal evidence, do you have scientific support for your worries? What is the probability of an approaching collapse? Or: Unless you can predict the exact timing of a collapse, isn't it wiser as "rational" individuals to keep on playing the market to maximize your profits? Luckily for them, Kennedy and Baruch were smart enough to realize that a bad omen was enough reason to leave the sinking ship.

Always Look for the Rats

> Clues are the most misunderstood part of detection. Novice detectives think it's about finding clues. But detective work is about *recognizing* clues. (Gran, 2011, p. 34, my emphasis)

Previously, I pointed out Kennedy's and Baruch's talent for observing and identifying bad omens. Understanding complex systems like crowds can benefit significantly from close observation and identification of simple signals that express the underlying dynamics. To elaborate on this idea, let us discuss rats.

Rats have a bad reputation, as expressed in common wisdom and expressions. For example, the expression "Like rats fleeing a sinking ship" suggests that the least loyal, the least trustworthy, and the most desperate are those first to abandon a failing endeavor. This does not do justice to the rats. It has been reported that they will show helping behavior toward a soaked conspecific (Sato et al., 2015). Human beings are prone to dismiss the moral and intelligent behavior of non-human organisms. Rats are treacherous and donkeys are stupid. But you only need to read Shakespeare to find all these traits among human beings and perhaps better appreciate the cognitive and moral behavior of non-human organisms.

The reason rats are the first to abandon a sinking ship is simple and trivial. Rats usually hide in the spaces first exposed to flooding. When the ship starts to take water, they are the first to notice it and head for higher ground. As Joseph Reinemann explained,[8] "As a result, rats suddenly appearing out in the open has long been used by crew and passengers alike as an early indicator of a leak in the hull in areas that may not be being closely monitored."

The lesson we should take from the rats is that, whenever we seek informative signs, we should seek them among those who are the first to pay the price for whatever is going wrong, as they are the most vulnerable. The shoeshine boy, who took a loan from the local loan shark to play on the stock market, was the one on the lowest decks of the ship. Those who do not have a stake in the game and might pay a painful price for misconduct should be ignored, regardless of their high status and impressive maneuvers. It is better to trust the rats than the sharks and the wolves of Wall Street.

[8] https://www.fascinatingearth.com/node/769.

A second lesson is that important and informative signs may fly "under the radar." Otherwise, everyone would have used them, and they would have lost their precious value. Betting against the crowd is possible if we can transcend our human biases of wishful thinking, over-optimism, and lack of criticism. Combining a deep scientific understanding of crowd dynamics and high sensitivity to "trivial" signs may be a good recipe for the individual to bet against the crowd.

In Sum ...

Kennedy and Baruch are famous examples of people who dared to bet against the crowd. I analyzed these examples to show that betting against the crowd is possible and sometimes necessary. The most general insight from the shoeshine anecdote is that profound simplicity may guide us in identifying bad omens that signal a tipping point. Exponential and potentially unsustainable growth may be powered by a positive feedback loop, pushing the system upward. If this growth cannot be supported at some point, it is likely to reach a dead end. Signs indicating the approaching bottom level of the pyramid are enough to take the "non-stupid" decision of leaving the sinking ship in good time.

Beyond the sophistication of predictive models with their numerous parameters and variables, there is a profound simplicity. Exponential growth may be unsustainable given real-world constraints (i.e., limited resources). With its greed, the madding crowd is involved in a positive feedback loop that nurtures itself and generates exponential or super-exponential growth. Simple signs indicating that the bottom of the pyramid is close may be enough to make the right decision.

Let me conclude with a personal anecdote. A talented data scientist that I know got a tempting business offer. When he told me about the proposal and the interaction with the businessman who offered it, I advised him not to take it up. The proposal was appealing, and the scientist questioned the scientific grounds for my advice. Whenever I pointed to "bad omens," he tried to look for more positively biased interpretations, questioning the non-scientific grounds of my advice. I finally made my point by using a metaphor. "It smells bad," I told him, "and if it smells bad, then you don't need to know whether the smell comes from the carrion of a rotten cat or a rotten rat. The decision should be simple: distance yourself from the bad smell."

A bad smell is a clear sign, even if we cannot depict its source and explain it. My metaphor worked. In retrospect, we found out that the businessman with his "respected" resume was no more than a confidence trickster. *Scientism* may lead us astray no less than naïve theories. Choosing the right thing to do does not necessarily require deep understanding. Sometimes, identifying the relevant signs is enough. The individual within the crowd may bet against the crowd by combining a profound understanding of the crowd's dynamics with recognition of suitably informative signs.

Adopting the ideal of non-stupidity, she may ask herself, like our reflective frog, three important questions: First, where are we heading to? Second, are there any signs that should bother me? And third, do these signs indicate a painful future for me and others? Depending on her answers, she may bet against the crowd and hopefully avoid the pyramid collapsing on the heads of the poor slaves building it for some new pharaoh.

References

Bateson, G. (2000). *Steps to an ecology of mind: Collected essays in anthropology, psychiatry, evolution, and epistemology*. University of Chicago press.

Bevelin, P. (2007). *Seeking wisdom: From darwin to munger* (p. 328). PCA Publications LLC.

Cipolla, C. M. (2021). *The basic laws of human stupidity*. Doubleday.

Corden, C. (2019). A Mathematical Deconstruction of Pyramid Schemes. *Leicester Undergraduate Mathematical Journal*, 1.

Goux, J. J. (1990). *Symbolic economies: After Marx and Freud*. Cornell University Press.

Gran, S. (2011). *Claire DeWitt and the City of the Dead*. Mariner Books.

Harries-Jones, P. (1995). *A recursive vision: Ecological understanding and Gregory Bateson*. University of Toronto Press.

Rothchild, J. (1996). When the shoeshine boys talk stocks. *Fortune, 133*(7), 99–99.

Sato, N., Tan, L., Tate, K., & Okada, M. (2015). Rats demonstrate helping behavior toward a soaked conspecific. *Animal Cognition, 18*(5), 1039–1047.

Sornette, D. (2009). *Why stock markets crash: Critical events in complex financial systems*. Princeton University Press.

Varshney, L. R., & Sun, J. Z. (2013). Why do we perceive logarithmically? *Significance, 10*(1), 28–31.

West, G. (2018). *Scale: The universal laws of life, growth, and death in organisms, cities, and companies*. Penguin.

Whalen, R. J. (1993). *The founding father: The Story of Joseph Kennedy*. Regnery Gateway.

3

Entropy, Constraints, and Action: Tools for Short-Term Prediction

Introduction

> That's what people miss. Our lives aren't built on hours, days, weeks, months, or years ... Life, like Hell Week, is built on seconds that you must win, repeatedly. (Goggins, 2022, p. 89)

If I ask you whether you will live through the next second, your answer will probably be positive. Now, I ask you about the next minute, and your answer will be positive too. After all, what can happen in a minute? But you may be less certain if I extend the time frame to an hour, day, week, month, year, or five years. Who knows what can happen in the next five years or so? The decay in our certainty is exponential. From this simple observation, we learn that our *uncertainty* is sensitive to the time frame of our measurement. This sounds like a trivial observation. After all, you do not have to be a genius to understand this point, but it is less trivial to understand the big "Why" and what we should do about it, specifically when betting against the crowd. To address these nontrivial points, I would like to take you on a journey that starts with a lady and a tiger and concludes with a decorated navy SEAL and his worldly wisdom. This little journey may help us understand that the "human particle" has degrees of freedom and that the opportunity to act may exist on very short time scales.

Y. Neuman, *Betting Against the Crowd*, https://doi.org/10.1007/978-3-031-52019-8_3

The Lady or the Tiger?

It always amazes me how great writers can grasp and represent deep scientific ideas in their short stories and novels. One of these stories is "The Lady, or the Tiger?" written by Frank Stockton in 1882 (Stockton, 1882). The story describes a sadistic king who developed his own unique form of *trial by ordeal*. Trial by ordeal was a kind of judicial practice in which a physically or spiritually challenging test determines the person's guilt or innocence. For instance, suppose a woman was suspected of witchcraft in the Middle Ages. To decide whether she was indeed a witch, she would be thrown into the river, and if, by God's mercy, she was saved from drowning, then her innocence would be considered proved. In retrospect, and from a scientific and moral perspective, the idea of trial by ordeal seems both stupid and perverse, as it is not a valid measurement procedure for innocence or guilt. A poor woman who does not know how to swim would drown despite her innocence, whereas a witch capable of flying on a broomstick and swimming like a fish would survive the test despite her satanic affiliations.

In *The Lady, or the Tiger,* we learn about a particularly sadistic form of trial by ordeal. It involves the accused being brought into a public arena. There, he is forced to choose between two closed doors. Behind one is a lady, selected by the king as a reward for the accused's innocence. Behind the other, a ferocious and hungry tiger waiting to enjoy a good meal and hence prove the individual's guilt. There is no way to discern the door of innocence from the door of guilt. The individual faces total uncertainty. He does not know which door to open as there is no clue as to what is hidden behind the doors. In the long run, if you could repeatedly "play" this game, there would be a fifty-fifty chance of meeting either the tiger or the lady. The king's sadistic pleasure probably comes from playing with the poor accused like a flipping coin and observing his horror while faced with the choice between the two doors. Those who strive for a strong leader, meaning a dictator, should read the story and remember that whenever there is a king, there is always the possibility that one day they will personally face two doors.

If you think about it, imagining ourselves as coins thrown by Lady Fortuna is a depressing idea. Without control over our lives, we acquire a kind of learned helplessness, like Pavlov's dogs. Freedom of choice is an old idea that has been debated for generations. I will not try to resolve it here, but I would just say that living out our lives with no sense of control, either true or imagined, is a depressing idea. Indeed, the horror we feel in reading this story results from the idea of facing total uncertainty, and hence a lack of *predictability* and *control* over the situation. No skill, statistics, or scientific knowledge can help the king's poor victim choose the right door. Lady Fortuna fully determines the person's destiny. Interestingly, Claude Shannon precisely conceptualized this uncertainty and developed a measure of uncertainty known as Shannon information entropy.

How to Measure Uncertainty

Like many great ideas, Shannon's entropy is simple, but hides conceptual depth worth examining. Let us start with a simple explanation. According to Shannon, uncertainty is a function of surprise, and surprise is a function of probability. Technically, Shannon's information entropy measures the *average* level of "informativeness" or surprise within a given *distribution* of outcomes. I explain this idea to my students using a simple example. Imagine residing in a country where the average height of men is 1.75 m. One day, your friend bursts into your office, informing you excitedly that he has just seen a man in the street who measures 2.50 m. How informative is this report? Is this "news"? Compare this situation to when your friend bursts into your office and informs you that he just saw someone 1.75 m tall in the street.

According to Shannon, the less probable an outcome, the more surprising and informative it becomes. Given the height distribution in this specific country, the probability of observing such a tall person is extremely low. So the surprise we feel in observing such a tall person is inversely proportional to the probability; the lower the probability, the greater the surprise and the more informative it is to report the "outcome" of such an observation. However, Shannon's entropy is not about a single outcome. It is about the average probabilities of outcomes. To explain this point, we must turn to combinatorics.

How many ways are there to arrange three objects A, B, and C? In fact, there are six ways:

1. ABC
2. ACB
3. BAC
4. BCA
5. CAB
6. CBA

Given N objects, there are $N! = N*(N-1)*...2*1$ ways of arranging them. These are the different *microstates* or possible *configurations* of the system. In our case, we have three objects. Therefore, $3*2*1 = 6$.

When analyzing the combinations for a large number N, we consider log N! Why? As explained before, this is simply to exploit the benefits of the logarithmic transformation when handling large numbers. For example, the number of possible configurations for ten objects is 3,628,800, while log 10! equals 6.56, which is an easier number to work with.

N datapoints can be grouped into C clusters of sizes $n_1, \ldots, n_C$. For example, imagine a country with ten citizens, where the distribution of wealth is such that 10% of the population is rich and the other 90% is poor. We then have $N = 10$, $n_1 = 1$, and $n_2 = 9$. Suppose Lady Fortuna is the only factor determining whether a given individual is rich or poor. She can produce different configurations of rich and poor individuals by spinning the wheel of luck. However, in principle, Lady Fortuna's distributions are constrained by combinatorics. She cannot generate infinitely many configurations of rich and poor people. Remember that 10% of the population is rich, and 90% is poor. How many ways are there to organize the data points (citizens) into the categories of rich and poor? The number of configurations for C clusters is

$$\frac{N!}{n1!*..*nc!}$$

In our case, this gives

$$\frac{10!}{1!*9!} = 10$$

which is a reasonable result when you think about it. Suppose you have ten individuals, and only one can be rich. In this case, we have ten possible configurations. In each configuration, one of the individuals enjoys wealth, while the nine others live in poverty. Assuming that a landscape of possibilities is defined by combinatorics alone, there are ten potential configurations.

The fact that among these ten individuals, Donald is rich, and the others are poor is just one possible configuration among the ten. Therefore, combinatorics defines the landscape of possibilities and has no interest in particulars. It tells us something interesting about the number of equivalent configurations (i.e., 1 rich and 9 poor), regardless of who is poor or rich. There are ten possible microstates in which one person is rich and the others are poor. This is all the information that combinatorics can give us. For Donald, it is highly important that he is rich rather than poor, but entropy is not about Donalds. It is about the abstract level of configurations showing relative stability over the erratic behavior of the particles, whether human or non-human.

The erratic level of the particles is the level at which we, as human beings, experience things from the first-person perspective. Whether John is rich or not is of no interest when measuring uncertainty, even though, for John, it may be a matter of life or death. Interestingly, it is in the erratic level of existence that our individuality is defined. However, science is about a higher level of abstraction where relative stability exists in a way that allows us to observe some order. In this context, there is an unbridgeable gap between our first-person perspective and the scientific perspective.

In a society with high inequality, we identify a Pareto-type distribution of wealth. This is a *low entropy distribution*, because wealth is not equally distributed. Regardless of those who are rich and poor, the distribution tells us something interesting about society. However, as Rovelli proposed, albeit in a different context, we pay too much attention to objects, when we should be thinking dynamically about change and relations. A person living in a country with a high level of inequality does not need economic indexes to teach him about the distribution of wealth. He knows that wealth is concentrated in the hands of the few. However, learning about *change*, such as social mobility and changes in social mobility, may be highly informative if we are interested in improving opportunities for talented individuals and removing less talented individuals from their unjustified power.

Let us return to our toy example. What happens if we believe half of the people are rich and the other half poor? In this case, we have 252 possible configurations. Notice that, when the distribution of rich and poor reached the maximal level of imbalance (i.e., one rich and nine poor), the number of potential configurations was smaller. So, it turns out that *the number of configurations is significantly higher when the distribution is more "homogenous,"* and the probability of observing one possible outcome (e.g., a rich individual) is the same as the probability of observing the complementary outcome (e.g., a poor individual).

This is an important point. The more balanced or homogenous a distribution, the more configurations it offers. This observation is closely associated with the idea of freedom. The more balanced and homogenous the distribution, the higher its Shannon entropy. The higher the entropy, the greater the "disorder" of the system. However, the greater the disorder, the more configurations. If you think about freedom in terms of your chances of being in one state or the other, then the more homogenous distribution of wealth increases your chances of being rich rather than poor. The appeal of middle-class society is the greater chance of having a good life. The great promise of the United States in the past was that it was not only the land of opportunity, but the land of a more probable bet on a good life as a part of a middle class.

Although chaos and disorder have gained a bad reputation, an acceptable dose of chaos is necessary for freedom. Whenever we encounter jealous and dogmatic idealists seeking either perfect order or perfect chaos, we should remind them that life exists on the boundary between order and chaos and that any attempt to impose total order or chaos (i.e., freedom) is a recipe for destruction.

Let me repeat the important observation that a more balanced distribution of particles means more possible configurations. What does this mean for our uncertainty? If we randomly pick a person from the first distribution, where 10% of the people are rich, is it more likely that he is poor or rich? The answer is clear. Most people are poor, so the odds favor the randomly picked person being poor. Let us measure the odds by dividing the probability of being "poor" by the probability of being "rich":

$$\frac{P(0.9)}{P(0.1)} = 9$$

The odds for the hypothesis that a randomly picked person is poor are higher than the odds for the hypothesis that the person is rich (i.e., 0.11). However, if we have the same proportion of rich and poor individuals, the odds are the same (i.e., 1), and we cannot improve our prediction beyond pure chance. We can predict better and improve our bet when the distribution is less homogenous. There is no way of gaining knowledge in a world of chaos or homogeneity. When the entropy is maximal, we are destined to live like particles floating randomly in a hot cup of tea, where the only order is that produced by Lady Fortuna.

The way in which the particles are distributed is directly associated with predictability and control. Moreover, it may tell us something about the *constraints* operating on the system, and even sometimes how these constraints change over time. This point is highly important for understanding the ideas presented in the book. Think, for example, about "The Lady, and the Tiger." Imagine that the sadistic king seeks to deprive his victims of any certainty. He therefore devises a perfectly balanced coin and tosses it to decide the location of the tiger and the lady. However, the doors are made so that the roaring of the tiger can just be heard. In this case, victims with highly sensitive hearing may have improved chances of avoiding the tiger. A deviation from a hypothetical distribution may be informative about regularities we seek to understand to gain prediction and control. Therefore, the maximal entropy of a system is maximal irreducibility in the sense that we cannot compress the mess of data to gain some efficiency and control.

A more homogenous distribution implies a greater number of possible configurations, which means higher uncertainty and less predictability. Shannon's entropy quantifies this degree of homogeneity. At this point, you can understand that entropy is not about the "order" of the world but primarily about *our ignorance* of the world, and only one possible way of quantifying it. For the vicious king in "The Lady, or the Tiger," there is complete certainty about where the lady and the tiger are hidden. For the poor victims of the king, it is the uncertainty that is complete. In the case of a homogenous or symmetric distribution, the uncertainty is maximal. According to this understanding, randomness corresponds to an even distribution. Therefore, randomness is an *assumption* that, through its *violation*, allows us to identify *regularities*. Whenever we observe a situation with a consistent deviation from randomness, this deviation can be conceptualized and indicative of *patterns of constraints* operating on the system. It's the difference between particles floating in a hot bath and particles with some predictive ability that can hopefully be exploited to control their destiny.

It should be emphasized that deciding which distribution is our point of reference is far from trivial. The distribution maximizing the entropy is something to be agreed upon before modeling the behavior and deciding how random it is, whether there is a significant deviation from randomness, and what this deviation tells us about the constraints operating on the system and driving it away from randomness. The simplest distribution to assume in a situation of complete uncertainty maximizes the entropy. When in complete ignorance, we cannot prefer one direction over another, so it's like playing dice with Lady Fortuna.

There is another nice way of understanding the idea of entropy. To explain this, let us assume that we have eight people. Seven of them are poor, and one is rich. In this case, the number of possible configurations is 8, and $\log_2(8) = 3$. What does this mean? Think about it in terms of the game twenty questions, where we ask binary questions, with answer "yess" or "no", and seek to identify an object through the answers we get. Here we have a representation of the eight individuals denoted by "R" for rich or "P" for poor:

R P P P P P P P

We wish to identify the rich subject and ask: Is (s)he located in the right four cells or the left four cells? The answer we get is: in the left four. Here is a new representation in which we have reduced our "problem space" by half:

R P P P

The next question asks whether the rich individual is located in the right two or the left two cells and the answer is: in the left. There now remain only two cells:

R P

Finally, we ask whether the subject is "hidden" in the right cell or the left cell. The answer is: in the left, and the individual we were seeking is finally revealed. How many questions did we have to ask? The answer is three, the base-2 log of our eight objects. In terms of Shannon entropy, we have 3 bits of information.

Let us understand how measuring the number of possible configurations is associated with Shannon's entropy. If we have a large number of data points, we represent these configurations on a logarithmic scale using

$$N\left(-\sum_{I=1}^{C}\frac{ni}{N}\log\frac{ni}{N}\right)$$

Using p(x) for n_i/N, the Shannon entropy is defined as

$$\mathrm{H(X)} = -\sum p(x) * \log p(x)$$

A nice explanation is this[1]: entropy counts the number of ways of categorizing N objects given a certain probability distribution. For example, if we repeatedly flip an unbiased coin, the probability of observing "heads" is assumed to be equal to the probability of observing "tails" ($p = 0.5$). The entropy is 1, the maximal entropy for a binary outcome with two possible values. Now, suppose you toss a coin and it falls on heads 75% of the time and on tails 25% of the time. In this case, the entropy has been reduced to 0.81. If Lady Fortuna were responsible for generating the outcomes of the coin tossing, then she would not prefer one outcome over the other.

[1] https://medium.com/towards-data-science/what-does-entropy-measure-an-intuitive-explanation-a7f7e5d16421.

Like Lady Justice, she is blind and does not prefer one outcome over the other. In the long run, after tossing the coin repeatedly many times, we should observe (approximately) equal probability for heads and tails and the maximal entropy score of 1. That is, in the *long run*, entropy is maximized. Why? Because these high-entropy configurations are more likely (assuming full ignorance). In contrast, if you observed the system's behavior diverging from the trajectory that Lady Fortuna and Lady Justice have sketched, then a reasonable conclusion is that some constraints operate on the system's behavior and form interesting patterns that may be used for prediction.

As time unfolds, the system configurations increase, and the entropy is maximized. Think about the distribution of men and women in the population. Certain periods (e.g., war) may be biased toward a larger proportion of women. However, the proportion of men and women is almost the same in the long run. More variables, interactions, and external influences can affect the outcome in longer time frames. Under certain assumptions, and as time unfolds and the size of the system increases, we should expect the system to behave more randomly and the entropy to increase. Moreover, as time goes on, the number of possible states and outcomes multiplies, and the information required to predict the future becomes increasingly complex. This is why long-term predictions tend to have higher entropy, making them more challenging and unreliable than short-term predictions.

Let me sum up. In contrast with some naïve interpretations, entropy does not measure order in the world but our degree of uncertainty or ignorance, given certain assumptions. It is, therefore, an *epistemic* rather than an *ontic* concept. It quantifies the way we see the world rather than the world itself. In addition, entropy is measured as an *average* over a "population" of instances *at the system's macro level (*i.e., *it is macroscopic).* Entropy does not refer to a single particle but to the population's average and relatively stable behavior. Moreover, entropy is technically simple but conceptually difficult to understand and apply. After all, the poor individual in "The Lady, or the Tiger" cannot improve his situation by understanding the concept of entropy. Any sophisticated understanding of uncertainty seems irrelevant to the individual particle, one of many composing the whole.

The Appropriate Dose of Chaos

Up to now, I might have created the false impression that uncertainty and surprise are "bad." The poor person in "The Lady, or the Tiger" did not enjoy the uncertainty. Indeed, this impression is far from being true. In her recent book "The Joy of Thinking," psychologist Zittoun (2023) points to the joy of surprise, from young children playing to scientists excited by new findings. Maybe we enjoy surprise because, regardless of its dark side, it is an inherent aspect of our freedom. The fact is that life does not work like a McDonald's factory, aiming to produce homogenous products with minimal errors. Life, including social life, is imbued with a variability that has its positive side. To better understand this point, it is important to emphasize the connection between entropy and variability.

The simplest and most general idea of entropy is presented by the physicist Gabriele Carcassi,[2] who suggests that entropy measures the *variability of a distribution*. When we observe variability, we know that some degree of freedom exists in the system, and we love freedom in the right amounts, as freedom entails opportunity. This point is repeatedly missed by individuals who advocate the totalitarian ideal of ultimate order and predictability.

An article was published in the New York Times under the title "In a heartbeat, predictability is worse than chaos." (Browne, 1989). It describes the work of Dr. Ary Goldberger, who made a counter-intuitive finding: variability in the rate at which our heart beats is necessary for adaptation and survival. To quote Dr. Goldberger, "The healthy heart dances, while the dying organ can merely march." Uncertainty is not therefore a "bad" thing in itself. It is necessary for life in the right amounts, in the right context, and with the right timing. Moreover, it suggests that, if there is a window of opportunity, it will be associated with uncertainty, a point we will discuss shortly.

The *Copernican Principle* states that we have no unique status in the Universe and that it is better to imagine ourselves like gas particles subject to the tyranny of Lady Fortuna. However, humans are reflective and tend to take things personally, even when it concerns randomness. As observed by Goggins (2022, p. 38): "Yet, the randomness of it all can feel so personal." Is there a place for hope, or are we particles floating helplessly in a bath of thermal equilibrium, condemned to follow the blind logic of Lady Fortuna?

The answer I would like to give involves two important concepts: *interactions* and *constraints*. I will start by discussing the idea of constraints. As time unfolds, any system will "strive" to maximize its entropy simply because a homogenous distribution of values is combinatorically more probable. However, constraints may be operating, specifically over very short time frames, significantly reducing the number of possible configurations. If that number is significantly reduced, the entropy decreases, our uncertainty decreases, and we may have better tools to predict and control our destiny.

---

[2] https://www.youtube.com/watch?v=-Rb868HKCo8&list=WL&index=16.

So, *short time frames may offer opportunities in a world governed by Lady Fortuna's logic* and by order. These pockets of uncertainty where constraints limit the system's degrees of freedom may be used for prediction and control. Later, I will discuss pockets of uncertainty, where the possibility of "breaking loose" and freeing ourselves from order means there are opportunities to be had. But first, I would like to discuss the benefits of short-term prediction and show how it corresponds to Goggins' idea of winning the next second. We will come to this when we have a better understanding of the important ideas of constraints and short-term prediction.

Understanding Constraints

To understand the idea of constraints and the way they may improve short-term prediction, let us assume that we have a time series of data points. For example, I observe a time series of stock prices and, at each step, I want to decide whether to buy another stock or sell my stock. Here is an imaginary example of such a time series, in which the numbers represent the monthly value of the stock:

$$S(t) = \{270, 7, 90, 33\}$$

As you can see, this time series ranges over very different values. Initially, the stock was at \$270, dropping to \$7 a month later. A powerful and interesting idea (Bandt & Pompe, 2002) is to convert the time series into *ordinal patterns*.

Let me explain. First, we decide on a block length that includes a certain length of numbers. This block is called the *embedding dimension* D. We segment the time series into overlapping blocks of length D and a delay τ that determines how many steps to the right we shift our block at each step. Let's see how this process works by selecting $D = 3$, and $\tau = 1$. Using our time series, we first break it into two blocks each containing three elements:

$$\text{Block 1} : 270, 7, 90$$
$$\text{Block 2} : 7, 90, 33$$

In a more visual representation:

270 7 90 33

270 **7 90 33**

The next step is to sort the elements in each block and represent them according to their relative magnitudes. For example, looking at the first block, viz.,

$$\text{Block 1} : 270, 7, 90$$

we are not interested in the absolute value of the stock, but in its relative value within the block. We note that 270 > 7 < 90 and represent this pattern using three numbers: 0, 1, 2. The first cell in the block has the highest value and is therefore scored "2". The next cell has the lowest value and is therefore scored "0". The third cell has an intermediate value and scores "1". The block is transformed as follows:

$$\{270, 7, 90\} \rightarrow \{2, 0, 1\}$$

It is thus mapped onto one of $D!$ permutations (denoted by π_i). These are the six possible *ordinal patterns*. As you can see, our discussion repeatedly returns to combinatorics, and later you will understand why. Meanwhile, for $D = 3$, there are six possible permutations/patterns/configurations:

$$\begin{aligned}
\pi_1 &= \{0, 1, 2\} \\
\pi_2 &= \{0, 2, 1\} \\
\pi_3 &= \{1, 0, 2\} \\
\pi_4 &= \{1, 2, 0\} \\
\pi_5 &= \{2, 0, 1\} \\
\pi_6 &= \{2, 1, 0\}
\end{aligned}$$

These permutations cover all possible configurations of the values within a window or block of three cells. This form of representation involves a significant reduction in our cognitive load. The range of possible results is huge if you examine the absolute stock price values within a window of three cells. However, when you represent each block of absolute values in terms of ordinal patterns (i.e., permutations), the enormous variety is reduced to just six patterns. Now, your mind only has to deal with six patterns which concern a small number of minimally connected units represented according to their relative rather than absolute values. Think of the enormous potential benefits of reducing the complexity of the time series to just a few patterns.

Adopting this form of representation, our time series can now be represented as a sequence of ordinal patterns. In our case,

$$\text{Block 1} : \{270, 7, 90\} \rightarrow \{2, 0, 1\}$$
$$\text{Block 2} : \{7, 90, 33\} \rightarrow \{0, 2, 1\}$$

The new sequence is

$$\pi_5 = \{2, 0, 1\}, \pi_2 = \{0, 2, 1\}$$

We see that the permutation π_2 follows the permutation π_5. We have a transition from π_5 to π_2. In other words, we observe a transition from one permutation to another. Consider the following question: Given a sequence of permutations, how many of the six different permutations can follow each permutation? Can permutation π_5 follow permutation π_1? Can permutation π_1 follow permutation π_3?

Naively, you might guess that, since we have six permutations (i.e., π_1 to π_6), each can be followed by any of the six permutations. So, permutation π_1 could be followed by permutation π_1, permutation π_2 by permutation π_6, and so on. However, *inherent constraints* are imposed on the *transition* from one permutation to another. Each of the six above-mentioned permutations may change to one of only *three* different permutations. This is an important point. It illustrates the way in which constraints reduce the number of configurations and may improve our predictions in the short run. Let me explain.

Constraints imposed on transitions from one permutation to the next do not exist for $D = 2$, but they do for higher-ordered embedding dimensions. For $D = 3$ and $\tau = 1$, *as in our example*, there are only *three legitimate transitions* for each permutation. For example, the second partition that we identified previously, viz., {7, 90, 33}, is mapped to $\pi = \{0, 2, 1\}$, where the order of the elements e_i is $e_1 < e_2 > e_3$.

Now, because the following partition/permutation (π_{N+1}) overlaps on the last two elements of the previous permutation π_N, the first two elements of the next permutation *must be ordered* according to $e_1 > e_2$, and the only remaining degree of freedom concerns the third element. This important fact comes about because we represented the time series as a sequence of overlapping ordinal patterns. Since the two consecutive blocks overlap, the choice of the third element is significantly limited, and so is the permutation pattern. For example, given our first permutation:

$$\pi_5 = \{2, 0, 1\}$$

the following permutation must be such that its first two elements e_1 and e_2 obey the constraint $e_1 < e_2$. How many of the six permutations obey this constraint? We note their first elements as follows:

$$\pi_1 = \{0, 1, 2\} e_1 < e_2$$
$$\pi_2 = \{0, 2, 1\} e_1 < e_2$$
$$\pi_3 = \{1, 0, 2\} e_1 > e_2$$
$$\pi_4 = \{1, 2, 0\} e_1 < e_2$$
$$\pi_5 = \{2, 0, 1\} e_1 > e_2$$
$$\pi_6 = \{2, 1, 0\} e_1 > e_2$$

We see that only the permutations π_1, π_2, and π_4 obey this constraint. This means that when, we observe permutation $\pi_5 = \{2, 0, 1\}$, only three permutation patterns can follow it.

It is important to realize that the constraints imposed on the transition from one permutation to the next significantly reduce the uncertainty associated with the next permutation, while naively you might have thought that six equally likely permutations could follow each permutation. Your ignorance/uncertainty would then have been quantified using the Shannon entropy as H(X) = 0.82. However, when the constraints are taken into account, there are *only three equally likely* outcomes, and the entropy is H(X) = 0.53. This reduction in uncertainty is a highly important cognitive resource (to be discussed further in the next section) because the *inherent* constraint substantially reduces the number of possible transitions from one permutation to the next, potentially improving the prediction of the π_{s+1} permutation in a *symbolic sequence of permutations* $\{\pi_s\}_{s = 1,\ldots,n}$. The list of legitimate transitions ($D = 3$, $\tau = 1$) from each permutation to the next is presented in Table 3.1.

Table 3.1 A list of legitimate transitions from a given permutation ($D = 3$, $\tau = 1$)

Permutation	Legitimate Transition To		
{0,1,2}	{0,1,2}	{0,2,1}	{1,2,0}
{0,2,1}	{1,0,2}	{2,0,1}	{2,1,0}
{1,0,2}	{0,1,2}	{0,2,1}	{1,2,0}
{1,2,0}	{1,0,2}	{2,0,1}	{2,1,0}
{2,0,1}	{0,1,2}	{0,2,1}	{1,2,0}
{2,1,0}	{1,0,2}	{2,0,1}	{2,1,0}

So far, we have discussed a very specific constraint resulting from our idea of representing the time series as a sequence of overlapping patterns. However, constraints may result from different sources and come in different flavors. In 1940, the great novelist John Steinbeck joined his friend, the biologist Edward Ricketts, on a voyage to the Sea of Cortez. Spanning over 4000 miles, this voyage triggered Steinbeck to reflect upon natural and human matters. As he explained, "Only in laziness can one achieve a state of contemplation" (Steinbeck, 1995, p. 151). This is not the usual, widely condemned type of laziness, but a mental state in which we free our mind for contemplation. Steinbeck's "laziness" could be usefully explained to academic bureaucrats and hard-headed individuals who believe only in hard labor and measurable performances.

In one of his reflections, Steinbeck discussed the enormous fertility of the sea hare *Tethys*. The number of eggs a single animal produces is staggering, but Steinbeck points out that "all this potential cannot, *must not*, become reality" (ibid. p. 111). *Constraints operate on potentials*, and one of these constraints is evolution. Breeding without constraints and under limited resources would lead to self-destruction. Interestingly, Steinbeck also reflected on how potentiality may be reduced when we go to the aggregate level of analysis:

"Statistically, the electron is free to go where it will be. But the destiny pattern of any aggregate, comprising uncountable billions of these same units, is fixed and certain, however much that inevitability may be slowed down" (ibid. p. 112.).

These are wonderful insights by a great writer. They may help us understand the human condition. In principle, a person may be considered as a free spirit with infinitely many degrees of freedom to explore all the possible trajectories of existence, like the character in Borges' short story *The Immortal* (Borges, 2000). Give us an infinite time to live, and we will explore all possible directions or life trajectories. These considerations can be extended to the idea of reincarnation. Give our "soul" the opportunity to return in different forms in an infinite universe, and we may become a king and a beggar, a human and a lizard, a Roman soldier and an American marine in Afghanistan.

But constraints must be acknowledged and considered when returning from Borges' fantasy to the worldly realm described by Steinbeck. As we are subject to constraints, it is perhaps somewhat paradoxical that we may free ourselves from the grasp of Lady Fortuna, who always has the upper hand in the long run. At this point, I would like to return to our specific example and explain why I have chosen $D = 3$.

On the Cognitive Importance of $D = 3$

We may now understand the importance of representing a time series using ordinal patternsspecifically those of embedding dimension $D = 3$.

If we chose $D = 1$, our analysis is meaningless. As we learned from Steinbeck, the isolated particle is a free spirit that cannot be tamed unless it is a part of an aggregate. If we chose $D = 2$, we convert our time series to a sequence whose elements are either $\{0, 1\}$ or $\{1, 0\}$. Keeping $\tau = 1$, in this case, $\{0, 1\}$ can change to either of $\{0, 1\}$ and $\{1, 0\}$, and likewise for $\{1, 0\}$, so there is no reduction in uncertainty.

When $D = 3$, constraints are applied, and we may start reducing the uncertainty. Moreover, for $D = 3$, the constraints imposed on the number of transitions from one pattern to the next are such that, keeping $\tau = 1$, the number of possible transitions is reduced *by half*, from six to three. Each permutation can change to only one of three different permutations. If $D = 4$, the number of legitimate transitions is only cut by 0.16. When $D = 5$, it is only cut by 0.04, and so on.

We see that the maximum relative reduction in uncertainty regarding the transition from one permutation to the next clearly occurs for transitions between permutations with length 3 (when $\tau = 1$). It seems that $D = 3$ is the most efficient representation for reducing the uncertainty about the next permutation in the sequence.

Finally, $D = 3$ is the embedding dimension where the number of possible transitions from one permutation type to the next is within the 'magical number' (i.e., 7 plus or minus 2). The magical number seven, plus or minus two, refers to a concept identified by cognitive psychologist George A. Miller (1957). Miller suggested that the average number of objects a typical human being can hold in their working memory is seven, give or take two. This means that, for most people, their working memory capacity is typically between five and nine items. When we use $D = 3$, the number of possible transitions is limited to 6. The number of possible transitions is within our limited cognitive capacity.

We have now discussed one source of constraint inherent in the way we represent the time series. Each permutation can change to only one of three possible permutation types, and the uncertainty concerning the next permutation is reduced by half. This constraint may be used to predict the fourth value in a sequence of four measurements. Let us return to our times series:

$$S(t) = \{270, 7, 90, 33\}$$

We represented the first three values through the ordinal pattern {2, 0, 1}, and we know that the next overlapping pattern must be one of the following three permutations:

$$\{0, 1, 2\}\ \{0, 2, 1\}\ \{1, 2, 0\}$$

We can thus guess the fourth value in the time series and, more specifically, whether it will be *higher or lower than the previous value*. That is, we do not claim to predict the exact value of the fourth value of our time series, but just whether it will be higher or lower than the value in the third cell. This is of course important if we are considering selling our stock before a decrease in value. I will show how we may use this idea in one of the forthcoming chapters.

In two out of three cases, the permutation following {2, 0, 1} is such that the fourth cell is LOWER than the third cell. When the following permutation is {0, 1, 2}, the fourth cell is higher than the third cell, and when the following permutation is either {0, 2, 1} or {1, 2, 0}, it is lower. This understanding may improve our prediction. Let us assume that we observe the first three values {270, 7, 90} and are asked to predict whether the fourth value will be higher than the third. In this case, in a situation of complete ignorance, the chance that the fourth will be higher is the same as the chance that it will be lower ($p = 0.5$). However, if we take into account the constraints, we can guess that the chances are higher ($p = 0.66$) that the fourth value will be lower than the third value. The entropy, and therefore our ignorance, have been significantly reduced.

However, we can do better. In practice, the chances are determined by the *transition probabilities* from one permutation to the next—in the above example, from {2,0,1} to each of the three possible permutations. Identifying the transition probabilities from one permutation to the next may provide us with another important source of information for short-term prediction. Given that we now 'know' that each permutation can change to one of only three possible permutations, learning the transition probabilities associated with each possible transition may further improve our prediction of the fourth value in the series.

We have learned that real-world constraints on a time series may significantly improve our short-term prediction. For example, understanding a crowd motivated by greed, we may expect that a monotonic increasing sequence of values (i.e., {0, 1, 2}) will be followed by another. Why? Because when people observe a market on the rise, they jump on the wave to enjoy the profit. Greed and jealousy are two important social emotions that motivate a positive feedback loop, pushing the market upward. If we understanding herd behavior and the constraints operating on it, we can use this to join the herd and play against it when required. To conclude this disscussion, let us turn to the fluctuation theorem and David Goggins.

How a Navy SEAL Uses the Fluctuation Theorem

The fluctuation theorem is a way of quantifying the ratio between increases and decreases in entropy. As Wikipedia explains, "as the time or system size increases, the probability of observing an entropy production opposite to that dictated by the second law of thermodynamics decreases exponentially." As time unfolds, the increase in entropy is inevitable, but for very short periods, there are probabilistic *windows of opportunity* where the entropy may even spontaneously decrease. I won't use the fluctuation theorem in its scientific sense, but draw an analogy.

Suppose there are very short periods where constraints operate, and entropy can even be spontaneously reduced. This means that opportunities will exist on very short time scales, to which we must be carefully attuned. Here we come to David Goggins. As he describes in his autobiography, he was born into a family in which an abusive father violently terrorized his wife and children. Escaping with his mother, he grew up in poverty. His lack of self-confidence led him to become obese and he ended up working as a low-paid pest controller. He tried to join the Navy SEALs, one of the American Army's toughest and most demanding elite military units. He wanted to change his life and become the toughest man on Earth. After failing on two attempts, he was finally accepted and became a SEAL. Moreover, despite extreme physical and mental challenges, he became an elite athlete and even broke the Guinness record for pull-ups.

In his reflection on Hell Week—one of the most physically and mentally challenging phases of SEAL training – Goggins explains to us his own strategy for survival and excellence:

> That's what people miss. Our lives aren't built on hours, days, weeks, months, or years … Life, like Hell Week, is built on seconds that you must win repeatedly. (Goggins, 2022, p. 89)

This is a profound reflection. Indeed, it corresponds to our discussion above. There are very short time frames in which we can beat Lady Fortuna. But repeatedly winning every second is a demanding task that requires a considerable effort. However, sometimes, a decision made in a split second may be a matter of life or death, or indeed financial collapse or wealth, as illustrated by Kennedy and Baruch. Focusing on very short time frames is a strategy well-grounded in our scientific understanding of complex systems and our understanding of ourselves as finite and limited "particles." So just disregard the fortune tellers with their ungrounded predictions and attune to the moment-by-moment struggle to win over the Lady. Short-term prediction is possible, and so are many short-term deeds that can greatly impact our lives.

In the cinema, several film directors have used the trick of playing the movie in reverse or incorporating reverse footage for certain scenes. For example, "Memento" is a psychological thriller directed in 2000 by Christopher Nolan. The thriller tells the story of a man with anterograde amnesia trying to solve the mystery of his wife's murder. The film uses a unique narrative structure, alternating between color sequences moving forward in time and black-and-white sequences moving backward. In real life, we cannot play things in reverse. Life is irreversible. Symmetry breaking is, therefore, a hallmark of life in which some order is formed and entropy is released into the environment, increasing the general amount of entropy in the system. In the long run, entropy increases, but as we learned, we may have the upper hand over a very short time frame and limited islands of order. Choosing the relevant time frame for intervention is therefore crucial to our success.

In the Midst of Chaos, There Is Opportunity

In addition to the time frame, two additional factors, in fact two processes, are relevant to understanding Goggin's insight. The first is memory. Memory is the correlation of past and present events. As far as humans are concerned, it allows us to record the past, reflect on it, and use it to identify regularities. Think, for example, about the young Goggins striving to get through Hell Week. Pain, cold, hunger, and sleep deprivation may lead to the thought: "I can't do it." This natural reaction would imply the response: "Let's quit!" By registering such a thought in our mind, even for a fraction of a second, we may delay the automatic response in favor of a better thought-out choice of action. Reflecting on the thought in memory, we may ask ourselves, "Can't you do it?" and this time we may come up with the reply, "Yes, I can."

His experience of disturbing thoughts has led Goggins to develop his own personal counter-strategies. In fact, SEALs are specifically taught how to handle such thoughts. Memory is a crucial feature that allows us to go back in time to change the present. Having memory means that past patterns are registered in our minds. We use these patterns to understand the present and speculate about the future. Later, I will deal with long-term memory, trends, and the way our understanding of crowds may be enriched by our understanding of the past. Memory allows us to identify patterns that constrain the short-term future. Entropy is thereby reduced, and so is uncertainty. Identifying these patterns ahead of the madding crowd may give the individual an edge. One possible route to betting against the crowd is to identify patterns and respond flexibly and promptly. The crowd, as a mass of people, will be less quick to respond.

The second ingredient that I would like to discuss is noise. Perhaps surprisingly, noise, random fluctuations obeying the logic of Lady Fortuna, may be used as a valuable resource. Think about it as card shuffling that may change your destiny by opening up new opportunities. "Aladdin's Wonderful Lamp" is one example where characters find themselves elsewhere or experience transformative events due to chance. Injecting noise increases a system's entropy in a way that may *release it from being trapped in a specific low-entropy configuration.* As such, it may allow us to find a new and better trajectory. Winning the next second may involve exploiting order, as when we recall previous patterns and introduce noise into the system, identifying opportunities for increased chaos, to free ourselves from the restrictions of some fixed configuration. The intricate interplay of order and disorder is not oppositional but complementary. Both disorder and order may contribute to our understanding of crowds and the possibility of successfully betting against the crowd. This important idea will be discussed repeatedly throughout this book. So, once again, note that order and chaos are not mutually exclusive trajectories. They should be used in a balanced way, in the right doses and with the right timing.

In sum, (1) finding the relevant time frame for intervention and using (2) order and (3) noise to produce reversibility are three possible mechanisms that may help us exploit irreversibility and reversibility for our own benefit. Winning seconds is just one example with profound implications for our daily conduct. Here, we may learn an important lesson about collectives and the empowered individual.

Windows of Opportunity: Pockets of Order and Chaos

"It is we that are blind, not Fortune: because our eye is too dim to discern the mystery of her effects, we foolishly paint her blind, and hoodwink the providence of the Almighty." Thomas Browne, *Religio Medici.*

In this chapter, I emphasize several important points. The first is that our uncertainty can be conceptualized in terms of entropy. Order, characterized by low entropy configurations, may be used to guess a system's trajectory. High entropy configurations may be a source of freedom from a fixed point or an attractor. Moreover, I explained the importance of constraints operating on the system, specifically for the short run, and the way they can be used to gain control over a situation. Finally, I mentioned Goggin's insight and the importance of identifying short time scales for intervention. This will be dealt with in the next chapter when we discuss "Borkmann's Point."

What can we learn from this highly abstract and scientific discussion about crowds and their dynamics? What can we learn about the individual seeking his way within the crowd? Crowds are collectives of particles operating under certain constraints, channeling their behavior into low entropy and predictable trends. However, as complex systems, they may also present unpredictable behavior and frantic shifts. The individual may use this understanding to see ahead and avoid fallacies, as explained in Chap. 2. However, the constraints determining the crowd's behavior cannot guarantee our prediction. Any advantage we can gain comes from understanding our limitations, as explained in Chap. 2, and the highly important role played by chaos. Moreover, as time goes by and the size of the system increases, there are no simple approaches to understanding the way our uncertainty grows. With this arsenal of new thinking tools, we may move forward to the next chapter.

References

Bandt, C., & Pompe, B. (2002). Permutation entropy: A natural complexity measure for time series. *Physical Review Letters, 88*(17), 174102.

Borges, J. L. (2000). *The aleph and other stories*. Penguin.

Browne, M. W. (1989). In heartbeat, predictability is worse than chaos. *New York Times*. https://www.nytimes.com/1989/01/17/science/in-heartbeat-predictability-is-worse-than-chaos.html. Retrieved Sept 18, 2023.

Goggins, D. (2022). *Never finished*. Lioncrest Publishing.

Goggins, D. (2018). *Can't hurt me: Master your mind and defy the odds*. Lioncrest Publishing.

https://www.nytimes.com/1989/01/17/science/in-heartbeat-predictability-is-worse-than-chaos.html. Retrieved Sept. 18. 2023.

Miller, G. A. (1956). The magical number seven, plus or minus two: Some limits on our capacity for processing information. *Psychological Review, 63*(2), 81.
Steinbeck, J. (1995). *The log from the sea of Cortez*. Penguin Books.
Stockton, F. R. (1882). The lady, or the Tiger? *The Century Magazine, 25*(1), 83–86.
Zittoun, T. (2023). *The pleasure of thinking*. Cambridge University Press.

4

Information Thresholds: Navigating Predictive Boundaries

How Much Information Is Enough?

> "Never enough data," Ron said. "When will there be enough data? There is no boundary to it." "There is," said the Chilean. "Just that you need to discover it anew every time". (Meiran, 1990, p. 21)

In Chap. 2, I emphasized the importance of signs for betting against the crowd when a pattern is identified in the data. The pattern in that case was one of exponential growth, which is always limited in any realistic situation. An important question that was left open is: How much information do we need to make a non-stupid decision? The basic questions are how to avoid loss and pain for myself and those that matter to me and how much information is required to act in a non-stupid manner. Finding a shoeshine boy involved in the stock market may be a bad sign, but should we reach an important conclusion, such as getting out of the roaring market just on the basis of a single case? More generally, how much more information do we need to make an important decision? Should we interview a representative sample of shoeshine boys before reaching a decision?

Y. Neuman, *Betting Against the Crowd*, https://doi.org/10.1007/978-3-031-52019-8_4

These are difficult questions. The insightful citation from Meiran's novel (see above) presents the idea that there is no clear boundary to the amount of data required. It appears limitless. The other character in the novel suggests that there is a boundary, but one that one needs to discover anew every time. This is an important insight: boundaries should be discovered anew. They should be discovered anew because, in particular contexts, general recipes cannot be applied.

In this chapter, I would like to address the question "How much information do we need?" by first using an insight from the thriller "Borkmann's Point." The hero, Inspector Van Veeteren, recalls an officer he greatly admired. This officer/mentor shared with him an insight he called the Borkmann point:

"In every investigation, he maintained, there comes a point beyond which, we don't really need any more information. When we reach that point, we already know enough to solve the case by means of nothing more than some decent thinking." (Nesser, 2006, p. 232).

So, the Borkmann point is the point beyond which we need no more information in order to act. We already know enough to solve the case using clear thinking. The thriller provides no "protocol" for identifying the Borkmann point. A naïve scientific approach might advise us to measure the value of information. If the incoming information does not change our knowledge or prior beliefs (in Bayesian terms), we may then think we have reached the Borkmann point. But this suggestion suffers from a problem. How could we know that the value of the information has not changed and that we have reached a plateau? We need some criterion, a benchmark, or a gold standard to measure the value of information. For a detective, the ultimate criterion is knowing who committed the crime, but he doesn't know this when he tries to address the Borkmann challenge. Otherwise, there is no point in asking whether we have reached the Borkmann point…

The challenge is much greater when we are dealing with the behavior of crowds. This is not a context where we have a simple system and past measurements to help us determine the predictive power of shoeshine boys. So, let us go back to via *negativa*, and instead of trying to measure the value of information, ask about "time to clearance".

Time to Clearance

"Time to clearance" refers to the time it takes for a criminal case, like a homicide, to be solved or closed by law enforcement. (Pastia et al., 2017) show that one-quarter of homicide cases remained unresolved. Most cases (68%) are cleared within the first week (see Fig. 4.1): "From these data, it appeared that the longer a case went uncleared, the more likely it was to remain so." This does not mean that cases remaining unsolved within a critical window cannot and will not be resolved. The distribution has a very long tail; sometimes, cold cases that have not been cleared in the first decade find their resolution years later.

Another way to view this is shown in Fig. 4.2. What we see is that the majority of the cases are cleared within the first week. These are simple cases, such as homicide-suicide, where a jealous husband has murdered his poor wife and then committed suicide. In this case, clearance is trivial. It is the simplest case possible, at least for a detective. We also see that the clearance frequency is associated with the severity of the case: the most severe cases are those with the lowest frequency, as happens with earthquakes. In fact, their severity is defined by the difficulty in clearing them. Cases that cannot be solved within a reasonable time frame are considered difficult.

To understand how time is related to the number of cleared cases, I used the curve estimation procedure of IBM SPSS Statistics 29 to test three models: linear, power, and exponential (see Fig. 4.3).

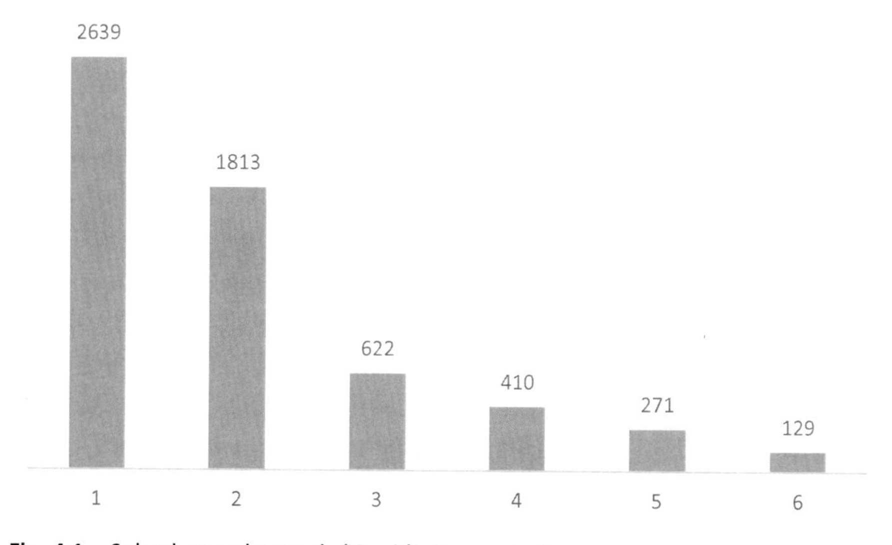

Fig. 4.1. Solved cases by week (X-axis). *Source* Author

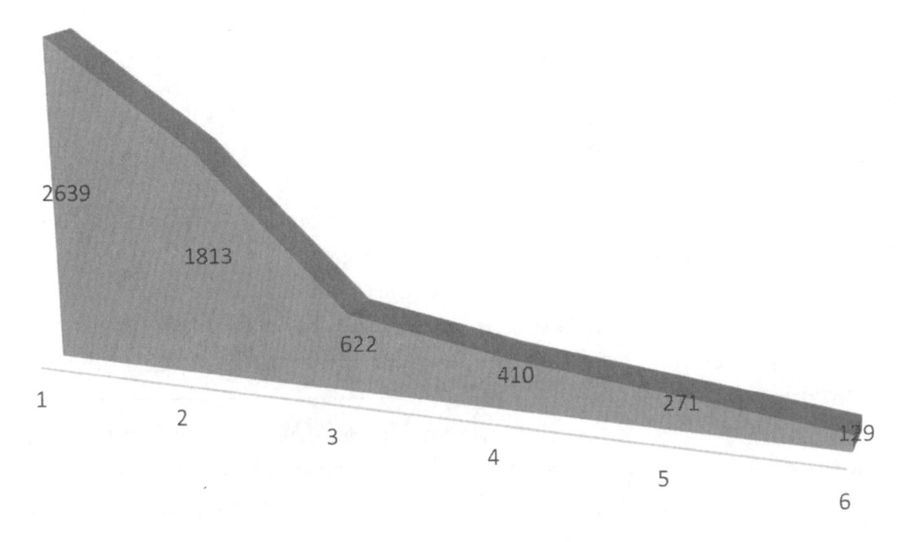

Fig. 4.2. The same data but with the emphasis on the ratio of part to whole. *Source* Author

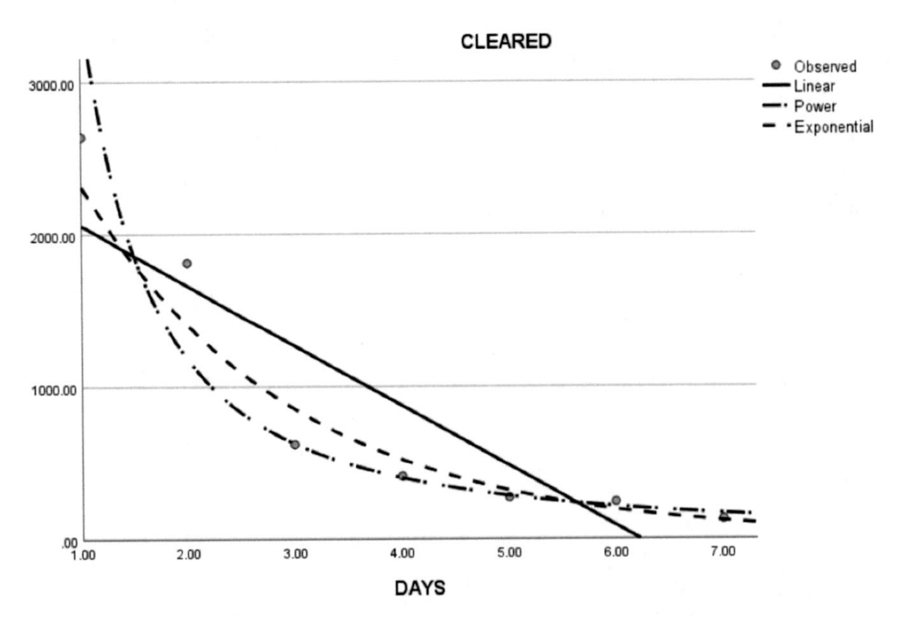

Fig. 4.3 Fitting curves to the data. *Source* Author

Using the adjusted R-squared statistic, which measures the explained variance, the power and the exponential curves show excellent model fits, with values 0.942 and 0.946, respectively.

Let me explain the difference between a power law *distribution* and an exponential *function*. Power laws are of great interest for describing distributions. They allow for the tail of the distribution to be extremely long or even infinite. When we look at cold cases, the murder of Irene Garza[1] was solved after 57 years, 7 months, and 22 days! Unlikely? Clearly not. The unbounded tail of a power law distribution always leaves room for a surprise. We can imagine future technologies that will help us to clear even older cases than Irene Gara's. Power law distributions are very important because they repeatedly turn up in social contexts regarding questions from wealth (e.g., the Pareto distribution) to wars. In one of the following chapters, I will discuss the dynamics of a conflict and the power law distribution of fatalities.

A power-law distribution is a statistical *distribution* in which the frequency of an event is inversely proportional to its size or magnitude. In other words, large events are less frequent, while small events are more common.

The severity of a case and its frequency are related in a well-defined manner (West, 2006). The more severe and extreme cases are less likely. When dealing with a time series of data, a power-law behavior implies a form of *long-range dependence*, where events at one point in time can influence events at a much later point in time.

In our case, the long tail is expressed in a non-linear decay rate that may result, for instance, from *evidentiary decay* (i.e., the decay of evidence).

The power law distribution, which describes how the *size* of events is related to their frequency, and the exponential decay function are related. In the context of crime and time to clearance, we know that most cases are solved, these being among the simple homicide cases, while a minority of cases stay unresolved, these being the more extreme cases.

The *exponential decay* of evidence is one of the main factors affecting the difficulty in clearing a case. Imagine a situation where the crime scene has not been analyzed within a week. In this case, the rain may have washed away important evidence, such as the remains of the perpetrator's DNA. *As time unfolds, it is not only the future that becomes uncertain but also the past.* This is why the first 48 h, which gave its name to the series "*The First 48,*" is considered to be such an important window of opportunity in forensic research. Time to clearance sets the time frame for collecting the relevant information. The exponential decay of information is very important, as it

[1] 10 Oldest Cold Cases Ever Solved—Oldest.org.

highlights the urgency of gaining the relevant information within a given *critical time window*. The decay of information is not linear. Information is not lost bit by bit over equal times. Most of it decays in the first period of the investigation.

The Exponential Decay of Information

Why is it that information decays exponentially? In the forensic context, there are simple explanations: the decay of evidence, witnesses' decaying memories, etc. However, we may return to entropy if we seek a more basic explanation. To recall, entropy measures the variability of a distribution, and when the distribution is homogenous, meaning that the probability of getting a certain outcome is equal across outcomes, then entropy and uncertainty are maximal.

There is a tricky aspect to this point that always confuses my students. The Shannon information entropy is maximal when the uncertainty is maximal. However, information exists whenever entropy is minimized, and some difference exists. Remember the story about the lady and the tiger? The idea of Shannon entropy suggests that information is maximized when the probability of meeting the lady or the tiger is the same ($p = 0.5$). Let us leave terminology aside for the moment. For mindful creatures like ourselves, order is produced when the entropy is reduced, and we can differentiate between categories. To describe the world efficiently, we must compress some information, and information compression is possible only when the entropy is not maximal. For example, human language is efficiently formed by using shorter words to describe more frequently observed objects and events. A world in which objects and events have exactly the same probability is a world where efficiency is impossible, a world of coin tossing governed by Lady Fortuna. Identifying differences, which are low entropy states, is what information is all about, at least at the most basic and abstract level of analysis.

The problem is that as the size of the system increases and time unfolds, there is a natural tendency for the maximization of entropy. Why? Because as I explained in Chap. 3, these are the more probable states from a purely combinatoric point of view. The second law of thermodynamics describes this trajectory in a very specific context. When structures and boundaries collapse, nothing holds the pieces together, and the second law has the upper hand. However, the same logic may be applied, at least analogously, to other contexts. For example, as time unfolds, DNA information in a low entropy state is "thermalized," deteriorates, and is lost. Continuing the same analogous way of thinking, the fluctuation theorem discussed earlier tells us that the probability of observing a trajectory that contradicts the expected one, in which the entropy increases exponentially, decays as a function of time or the size of the system. But why? Why exponentially? Let me present an answer in several steps.

Let us start by thinking about the natural decomposition process. When a living being, such as a human being, ends its life, its body starts decomposing. The brain is the first part to decompose because it is the hungriest organ (Raichle et al., 2002). The brain constitutes only 2% of our body weight, but it consumes a disproportional amount of energy (about 20%). We understand immediately that energy is not *evenly distributed among the body's organs and systems*. The most energetically demanding systems are the first to decompose. After death, with no active energy input to maintain the brain's specialized cell structure and function, the energy previously allocated to the brain begins to redistribute, whereupon the system shifts from a low entropy configuration to a higher one. Interestingly, this energy redistribution is probabilistic and obeys the principles of the *Boltzmann distribution*. Due to its high energy demand during life, the brain's energy state is relatively high compared to other tissues. As the entropy increases and the energy redistributes, the brain's energy state decreases more rapidly than that of other tissues. This leads to the relatively rapid decomposition of the brain compared to other parts of the body.

While the Boltzmann distribution is a powerful tool in physics, it is primarily used to describe the distribution of particles among different energy states. Applying it directly to the process of body decomposition would be a significant simplification and may not provide a comprehensive explanation. To draw an analogy, we may consider the various components of a deceased body as "energy states." Each component (e.g., the brain, muscles, organs) could be assigned an "energy level" based on factors like metabolic activity, cell complexity, and vulnerability to decomposition. Using this analogy, we may describe the decomposition process in terms of a simplified Boltzmann-like equation:

$$P(E) \propto e^{-\frac{E}{\kappa T}}$$

where

$P(E)$ Represents the "probability" or likelihood of a component decomposing,

E Represents the "energy level" of the component,

k Is the Boltzmann constant,
T Represents the "temperature" of the environment, which influences the decomposition rate.

The most energetically demanding systems are the first to "decompose." Their exponential decomposition is expressed as a cascade in which those disproportionately energy-demanding systems contribute most to the process. As we know, the surrounding temperature is crucial to the rate of decomposition of a corpse. This fact is expressed by T in the above equation.

So far, I have been using analogies. But why does information decay exponentially? The answer is that information is not equally distributed, and the more "energetically demanding" parts are the first to decompose, as happens with the brain. Think about the *Ebbinghaus forgetting curve*, which expresses the exponential decay of memory. The information absorbed by our brain is mostly meaningless. Think about how much irrelevant garbage you are exposed to daily in the media. How much of this information is valuable? 10%? 5%? 1%? I do not have the answer, but it is clear that the Pareto law also applies to this case. Storing all of this information in our brain would be an energetically demanding task, and the most efficient way of dealing with it is to throw most of it away in the early stages of cognitive processing. Underlying our existence, there is a deep logic regarding energy, and it covers various aspects of our life, from the decay of a corpse to the decay of information. Over time, the probability of a high-energy and low entropy decreases exponentially, which means that as time progresses, the likelihood of the system being in a high-energy state gradually decreases.

Now, let me present a deeper scientific explanation. We begin with the idea of configurations discussed previously and return to the example of rich and poor people. Assume that we have 4 particles. These are human particles. Each particle can be in one of two energy states, E1 and E2, corresponding to their wealth. E1 means one is rich, and E2 means one is poor. These energy levels have values. For simplicity, I assume that the energy level of E1 = 2 and the energy level of E2 = 1. This is a highly simplified example. When we discussed the demanding brain, we learned that it constitutes 2% of our body weight but consumes 20% of our energy intake. In my example, being rich means having twice as much energy as being poor. No more. Given 4 human particles, there are four possible configurations or microstates:

1. All particles are E1 (rich).
2. Three particles are in E1, and one in E2.
3. Two particles are in E1, and two in E2.

4. One particle is in E1, and three are in E2.
5. All particles are in E2 (poor).

The energy levels associated with each microstate are therefore:

1. $4 \times 2 = 8$
2. $3 \times 2 + 1 = 7$
3. $2 \times 2 + 2 \times 1 = 6$
4. $2 + 3 \times 1 = 5$
5. $4 \times 1 = 4$

We find that each of the possible configurations has a different energy level. The specific configuration of particles and their respective energy levels determines the energy level of each microstate.

The next step is to calculate the probability of each microstate from the Boltzmann distribution. For simplicity, we assume $kT = 1$, so the probabilities are as follows:

1. $e^{-8} = 0.000335$
2. $e^{-7} = 0.000911$
3. $e^{-6} = 0.002478$
4. $e^{-5} = 0.006737$
5. $e^{-4} = 0.018315$

These probabilities should be normalized so that their sum is equal to unity, but even now, we can see that the probability for a system to be in its highest energy state, when all the particles are rich, is the lowest.

There is a nice explanation here as to why societies failed when they tried to force full equality. Forcing full equality is like forcing a high-energy state on the various particles in the game of wealth when the energy resources in the system are finite. However, one should also notice that the situation in which all particles are poor is the most likely. So, why do we not see societies where *all* citizens are equally poor? The answer is that the process is more complicated. A system with no differences in energy between its subsystems cannot generate work. A society where everyone is the same is a frozen, isolated island where nothing can be created. This is why the extreme tails of energy levels don't apply to living systems, where the distribution must involve some variability.

In our present context, we can progress even with such over-simplified assumptions by plotting the probabilities of the microstates on the Y-axis and discrete time points on the X-axis. The probabilities have been multiplied by 10 000 to ease visualization. Proceeding in this way, we discover an exponential decay pattern (Fig. 4.4).

A system in a high-energy state would decay exponentially into a low-energy state. A body with its energetically demanding brain has its own complex decay process when it dies. However, the brain is the first to go among the organs as it is in a high-energy state.

Let us weave together all the threads we have discussed so far:

1. **High-Energy States and Low Shannon Entropy**

High-energy states in a system often correspond to low-probability states. There are fewer microstates associated with them. In terms of Shannon entropy, this corresponds to low entropy (uncertainty).

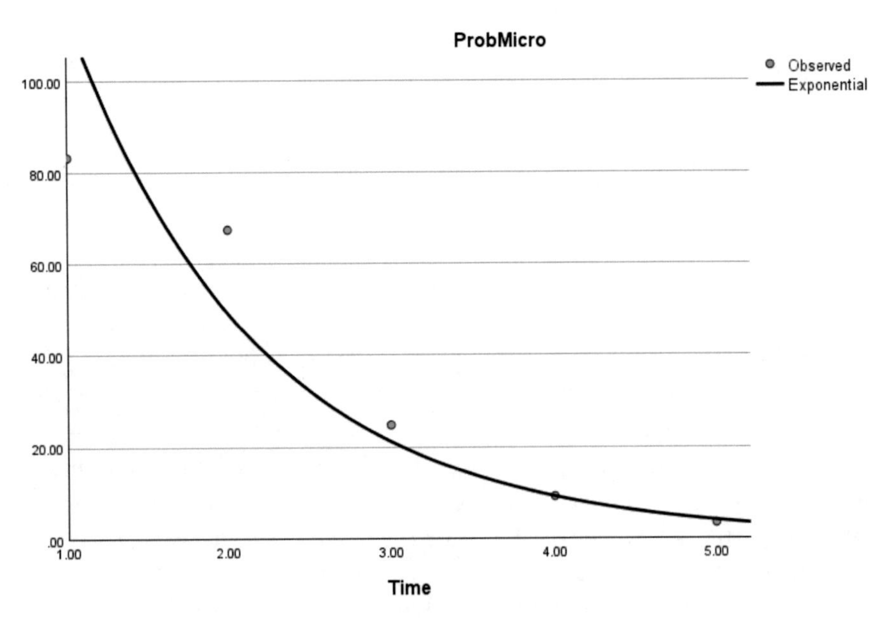

Fig. 4.4 Decay of probability as a function of time. *Source* Author

2. **Transition to Low-Energy States and Increased Shannon Entropy**

As time unfolds, the system tends to move towards states with lower energy (due to the probabilistic nature of the particles and the second law of thermodynamics). These low-energy states often have a higher probability of occurring and therefore higher Shannon entropy (higher uncertainty).

3. **Exponential Decay and Information Loss**

The exponential decay of the high energy to low entropy states corresponds to a decay in information because the system becomes more disordered. It follows the decay pattern suggested by the Boltzmann distribution.

So, How Much Information Do We Need?

Time to clearance does not set the time frame "beyond which, we do not really need any more information." It points to an interesting decay function where relevant information might deteriorate quickly as time unfolds.[2] This decay function may differ in different domains, as expressed by the idea of a critical window.

A window of opportunity or a *critical window* appears in domains ranging from development to emergency medicine. In emergency medicine, it is common to talk about the "golden hour" for saving the life of a trauma victim. The golden hour is a rule of thumb and should not be taken literally. The critical time for saving an individual is contextual and depends on factors ranging from the severity of the trauma to the individual's age. However, in the case of a heart stoppage, the death rate is higher than 50% after the first 3 min, and for massive bleeding, it is 30 min. The function associating time and death rate for cardiac arrest typically follows an exponential decay curve. As time goes by, the chance for a happy ending rapidly decreases.

This understanding may have clear implications for anyone betting against the crowd. Gaining an edge or a profit from a stock involves selling it at a higher price than you bought it. Here, timing is critical, as for the golden hour in emergency medicine or the first 48 h in a forensic investigation.

[2] What deteriorates is not necessarily information, but the human resources allocated to the mission. However, I am focussing here on the deterioration of evidence.

The efficient market hypothesis (EMH) suggests that financial markets are "efficient" in incorporating and reflecting all available information into the prices of securities. According to this hypothesis, it is extremely difficult, if not impossible, for an investor to consistently outperform the market by exploiting publicly available information.

For example, assume that a leading pharmaceutical company announcing a new drug for treating Alzheimer's disease has successfully passed the clinical trials. Once this announcement has been released, the stock price should immediately rise. There is no edge for those trying to beat the market as the information becomes available to all players. The rate of information decay is such that, unless you are the first to respond, you may be the last to earn or, in the case of collapse, among the many who might lose. This is an important point that will be discussed further. To bet against the crowd, one must be sensitive to the time scale available for action and the function describing the information decay. Not all time scales are born equal, and for the individual seeking to bet against the crowd, it is important to identify the critical time period for taking action.

Let me summarize the argument. When dealing with a power law distribution, where a minority of the cases are responsible for the most extreme events, our attention should be given to the disproportional price of these events. In this context, we should be sensitive to the exponential decay of the available information. This decay function determines the critical window in which we may successfully intervene.

At this juncture, I would like to present the idea that the Borkmann point is not determined by some hidden and intrinsic value of the information but by the *critical window* in which it appears and by its *decay function.* The point beyond which we do not need more information is a point, or more accurately, a region, beyond which information decays to the level where it may become irrelevant or extremely difficult to obtain. According to the EMH, beating the market is impossible. Still, innovative tech companies such as Renaissance, founded by the legendary Jim Simons (Zuckerman, 2019), did manage to do it. They did it first by identifying an extremely short time period where they had an edge, the region beyond which there was no edge to be had. The EMH is not a divine rule but a general observation made under certain conditions, and this is why the possibility of beating the market or betting against the crowd has been proved beyond doubt.

The shoeshine boy incident could have led Kennedy and Baruch to conclude that the market was becoming too efficient to the point of saturation and that there was no edge to be had in continuing the game. When approaching the region where no edge can be gained, one is destined to be led by the blind collective. Leaving the market means getting an edge by avoiding the domination of the collective with its frantic rush into the abyss. In the context of the stock market and the spread of information through media and rumors, the critical time has been shortened to an extent that Baruch and Kennedy might have considered too risky.

The bottom line is that the Borkmann point may be determined by a domain-specific critical window and the specific function of exponential information decay. Gaining an edge over a madding crowd may have a similar logic that does not necessarily accord with our psychological time and wishful thinking.

In sum, we began the chapter by asking how much information we need to avoid stupidity. My answer was that we should use a different approach in the context of crowds. This approach involves gathering all possible information in the critical window set by the context. If you are a detective, you know you have a limited time window for gaining evidence. If you are a trader, your time scale may be much shorter than the detective's, especially in a digital era where information spreads at unprecedented speed. Therefore, the question is not how much information I need to be able to decide but the time window available for collecting information before losing a possible edge or paying a catastrophic price. Betting against the crowd does not always mean that we have to have any special ability to predict the future, only that we know how to identify the critical time period of the game and the exponential decay function of information in the specific context. The way our failure to correctly predict the future characterizes our misunderstanding of crowds is discussed in the next section.

Hic sunt dracones

> In accordance with the plan laid down, we proceed to the consideration of the follies into which men have been led by their eager desire to pierce the thick darkness of futurity. (Mackay, 1852, p. 281)

We have previously pointed to our shortcomings in predicting the future (Chap. 2). Those who believed that the stock market would increase forever were wrong. Telling what the future will bring is always a problem. This is why serious people try to avoid it as much as possible, leaving the task to those professions with less scientific methods, such as fortune-tellers using cards. Does it mean that trends do not exist? The answer is negative. Trends do exist, but maybe our fascination with trends is the problem. Freedman (2010) explains the streetlight effect—searching for something only where it is easiest to look—and points to the scientific obsession with measurement. Of course, science is all about measurements, but measurements have their limitations. As will be explained in the third section of this chapter, we all live in Plato's cave, observing low-level signals from a much more complex reality. Observing the shadows of social behavior may be an excellent approach as long as the crowd does not surprise us with its long tail behavior.

However, there is a point for modest optimism: crowds may not be fully predictable, but our ignorance is. What does this tell us? That we may learn about the potential extent of our ignorance, at least for short-term extrapolations. To illustrate this, I have shown the first twelve measurements of the Dow Jones index in Fig. 4.5 and tried to identify the function that best describes them. As you can see, the increase in the DJ is exponential. It is almost perfectly described by an exponential regression model (adjusted $R^2 = 0.97$).

Now, imagine a brilliant trader observing the exponential increase of the stock market in the first year after it has been established. The trader is so impressed by the model's performance that he extrapolates it to the next year, i.e., the next 12 data points. The results are shown in Fig. 4.6.

You can see that from month 13 onward, there is an increasing gap between the extrapolation and the actual performance of the Dow Jones. The growth has turned from exponential to logistic. The simple and almost perfect exponential model failed to predict the future and signaled that the greedy crowd's behavior was more complex than expected. If we plot the absolute error of our prediction, which I have transformed for the sake of visualization in Fig. 4.7, we observe a perfectly straight line that can be modeled by linear

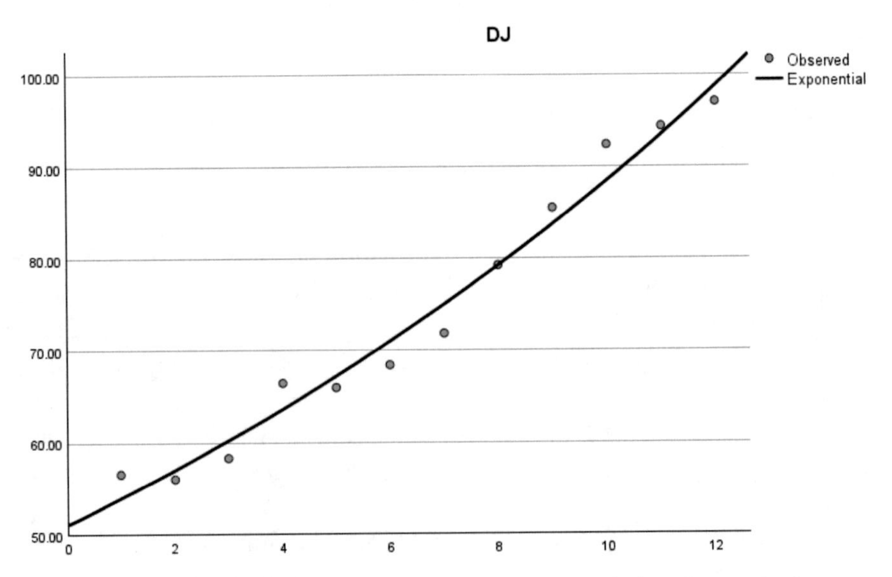

Fig. 4.5 Fitting an exponential model to the Dow Jones index. *Source* Author

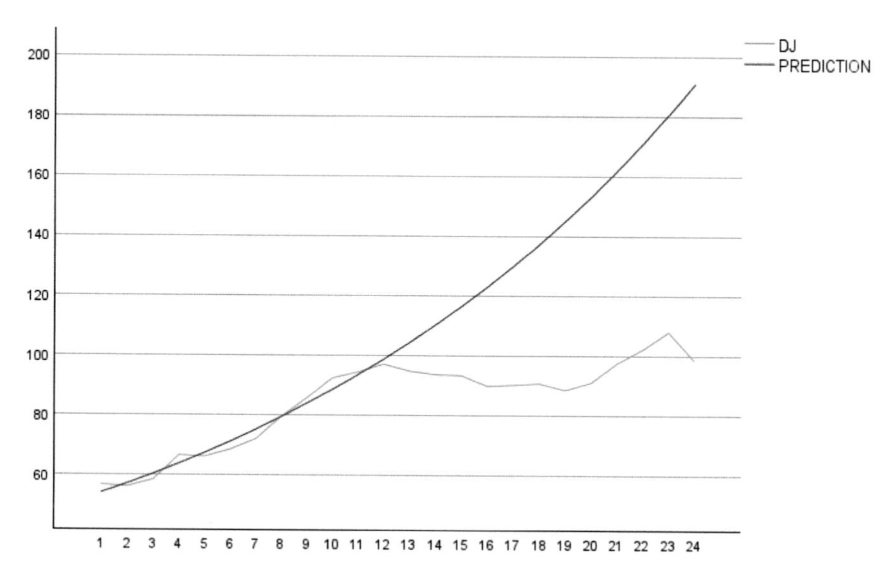

Fig. 4.6 Extrapolation of the exponential model. *Source* Author

regression (adjusted $R^2 = 0.98$). The errors grow linearly. For each month, I know that the error in predicting the exponential trend will increase by a fixed level of punishment.

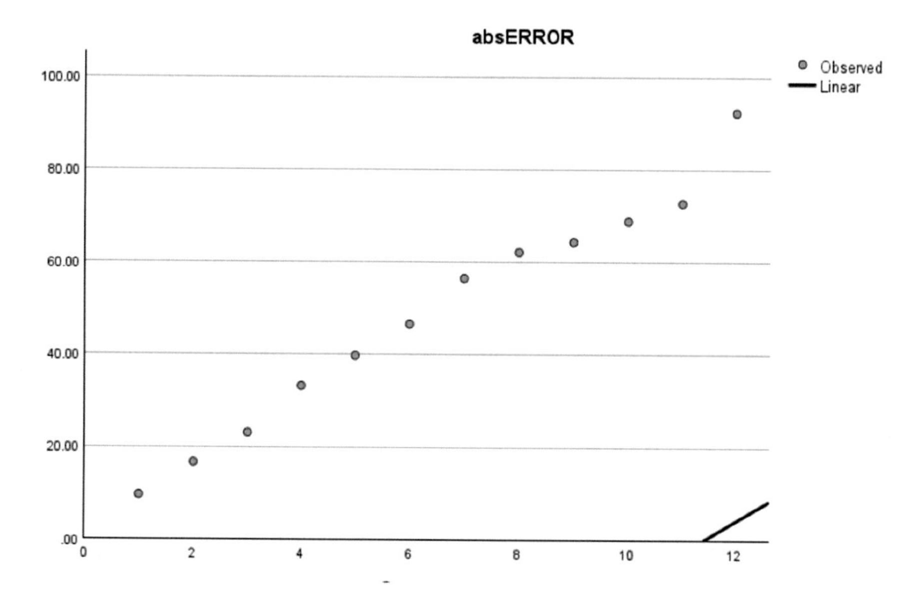

Fig. 4.7 Plotting the absolute error of the model. *Source* Author

The territory marked by the difference between the extrapolation and the actual DJ index is dangerous, and we must avoid this region because *hic sunt dracones*. This term was used in ancient times to describe the dangerous territories governed by dragons. The dangerous territory we are dealing with in this book is the territory of the crowd behavior, populated by unpredictability rather than by dragons.

Does this mean that trends do not exist and that, being afraid of dragons and uncertainty, we should avoid any extrapolation? The answer is negative. The reader should avoid any strawman interpretation of my arguments. Of course, we may bet in the long run and gain a significant profit from investing in the stock market. This is an excellent choice if we are not shoeshine boys or investors who put too much of our wealth into the game.

Is there a fixed time window where we can make safe predictions? Do we know how our errors will behave when extrapolated? As long as it concerns crowds, we may be safe if we identify some trend. Still, as the behavior of crowds involves a strong component of unpredictability, we are prone to get hurt by extreme events, such as the collapse of the stock market or the outburst of a war, as will be discussed in Chap. 9.

Psychologically speaking, one major mistake we are prone to make is believing in the persistence of trends. We desperately seek order, even in places where it does not really exist, and if it exists, the crowd holds it close to its heart despite its temporary and erratic existence. The situation is much worse when dealing with a group of people like ourselves. Our familiarity with other human beings can be misleading. The fact that we are surrounded by human beings is not a guarantee of understanding, not to mention pointwise prediction. Remember Dogville. At what point does the group of good American citizens turn into a cruel and treacherous mob? At what point does exponential growth turn into a logistic curve? If there is a lesson to be learned from the discussion so far, it is that trends are indeed important, but no less important is the understanding that we are extremely limited in identifying where a collective of human beings arrives at the point of phase transition.

Out of Plato's Cave

One of the most famous parables in the history of philosophy is Plato's cave. The parable goes as follows. Imagine a cave where prisoners have been held captive for their entire lives. They are held in place, unable to turn their heads, and can only see the wall in front of them. Behind them is a fire, and between the prisoners and the fire, there is a parapet where puppeteers can walk. The puppeteers cast shadows on the wall in front of the prisoners by holding up various objects before the fire. The prisoners can only see these shadows on the wall, and believe that they constitute the only reality.

This is only the beginning of Plato's insightful story. It appeals to every human being who has ever wondered whether what we see and understand really represents the truth or whether we are destined to live in a cave filled with illusions. However, there is another interesting point in the story. The shadows are low-dimensional projections of a much more complex reality. When we observed the fluctuations in the price of the Dow Jones, we observed a low-dimensional signal, which is the value of the Dow Jones across time. This signal is like the shadows in the cave. It is a projection of a much more complex situation in which puppeteers buy and sell stock according to dynamics that we do not really understand, at least when it comes to the tipping point of collapse.

How can we escape Plato's cave, assuming that our mind can only digest simple representations of a complex reality? What does it mean for our understanding of crowds and the dangerous trajectories they may follow? In the first section of this chapter, I explained that the question "How much information do we need?" might be misleading. In fact, we have no way of answering this question. Instead, we may think about the time frame available to us before the information decays in such a way that any advantage is lost.

As I conclude this chapter, we can begin to understand the way in which the Hamas terrorists were able to break into Israel and perpetrate a massacre. The Israeli intelligence was blind to the plan. However, even worse, as the terrorists broke into Israel, they blinded the military observation systems that protected the fence, entered the villages, and began to carry out their massacre with impunity. With such a short distance between armed terrorists and peaceful villages, it could have been expected that the *redundancy* of the backup system would be a default strategy, given the exponential decay of information over extremely short times. The critical window was too short. The blindness of the intelligence agencies was an expression of the *hic sunt dracones* problem. No one saw any real reason for concern when they examined the trend preceding the extreme event.

Russell (2001) told us an insightful story about the problem of induction. But let me tell the story with my own twist. The story involves a chicken on a farm. This chicken is fed daily at the same time, and it observes that, every time the farmer appears, it leads to food. The chicken makes an inductive inference: "The appearance of the farmer leads to food." cognized past trend. Moreover, the farmer increases the number of grains he gives the chicken. The chicken may become optimistic, like the happy Chinese mentioned in

Chap. 1. It begins to recognize a trend, and a good one: food is not only given regularly but also increases in quantity! However, one day, the chicken meets an unexpected fate—it is slaughtered. From the chicken's perspective, this event was entirely unforeseen and shocking. The inductive inference that the farmer's appearance always leads to food was shattered in one unforeseen, catastrophic event. As humans, we repeatedly make the same mistake. It is remarkable how similar we are to the poor chicken. When observing human crowds, we are surprised again and again. Could we imagine a chicken wise enough to avoid this fallacy?

References

Freedman, D. H. (2010). Why scientific studies are so often wrong: The streetlight effect. *Discover Magazine, 26*, 1–4.

Mackay, C. (1852). *Memoirs of extraordinary popular delusions and the madness of the crowds*. Richard Bentley.

Meiran, R. (1990). *South of Antarctica*. Keter. (in Hebrew).

Nesser, H. (2006). *Borkmann's point: An inspector van veeteren mystery*. Vintage Crime/Black Lizard.

Pastia, C., Davies, G., & Wu, E. (2017). Factors influencing the probability of clearance and time to clearance of Canadian homicide cases, 1991–2011. *Homicide Studies, 21*(3), 199–218.

Raichle, M. E., & Gusnard, D. A. (2002). Appraising the brain's energy budget. *Proceedings of the National Academy of Sciences, 99*(16), 10237–10239.

Russell, B. (2001). *The problems of philosophy*. OUP Oxford.

West, B. J. (2016). *Simplifying complexity: Life is uncertain*. Bentham Science Publishers.

Zuckerman, G. (2019). *The man who solved the market*. Penguin.

5

Short-Term Fluctuations Versus Long-Range Trends: Unearthing Investment Opportunities

Introduction

> By all means, make forecasts – just don't believe them. (Makridakis et al., 2010, p. 88)

Betting against the crowd is usually a bad idea. This is not because of any wisdom that might be attributed to the crowd. The crowd's wisdom suggests that a group of individuals' collective opinions or decisions may be more accurate and insightful than those of any expert or individual within the group. However, like all general ideas in the social sciences, it is fundamentally wrong and uninformative. The wisdom of the crowd is not a fundamental law of physics, and the idea that the crowd is often (?) more accurate than the "individual" or the "expert" is of little value, because the "often" is irrelevant for cases where we observe a long tail. In such a case, the "general," usually expressed by the average, is irrelevant.

Moreover, the crowd's wisdom depends heavily on the non-linear interactions that may result in sub-optimal or super-optimal performance. A crowd may make better decisions than an expert (but which expert?) or awful decisions as compared with an individual or an expert. The question "When is the crowd's wisdom expressed?" is therefore best given the cryptic answer: it depends. Betting against the crowd is not necessarily betting against the crowd's wisdom, it is betting against the power of the crowd. Therefore, by all means join the herd and ride on the rising wave in times of peace and financial growth. This is simple advice every grandmother should know. Join

Y. Neuman, *Betting Against the Crowd*, https://doi.org/10.1007/978-3-031-52019-8_5

the winners rather than the losers. It is similar to investing in the S&P 500, one of the world's most widely followed stock market indices. How many "genius" investment gurus do you know who can repeatedly and significantly beat the S&P in the long run?

There is no doubt that the crowd has an enormous power over the individual. In contexts varying from politics to finance, the price of betting against the crowd may be too high in many cases, not because of the crowd's wisdom, but because of the power of the crowd. Moreover, betting against the crowd is a doomsday strategy that can seldom be operated, as it requires access to information that can give a clear edge. Access to such information is rare. After all, how many people can gain reliable forecasts of a falling stock market or privileged information about an approaching war?

The head of Israel's Mossad obtained just such a piece of privileged knowledge before October 1973. The reliable Egyptian spy informed him about a war planned against Israel. He even gave the exact date on which the war would start. Israel's leaders did not accept the warning, even in this unique case of highly accurate and privileged information. The end was a total surprise and a painful war, which Israel won but at a great price. The political and military crowd won in the debate with the head of the Mossad, and the country lost.

So, what is the point of betting against the crowd? The point is that, although betting against the crowd is not easy, this does not mean it is impossible. Sometimes, it may be beneficial, and sometimes, it is a moral obligation.

Buy-Low-Sell-High (BLSH)

In this chapter, I want to show that being contrarian is possible if you are creative enough to implement some of the ideas presented in the previous chapters. The presentation intends to be practical. It aims to show how to play a contrarian approach through the heuristic known as buy low and sell high. This approach is appealing due to its simplicity. It involves purchasing assets (such as stocks, real estate, or other investments) at a relatively low price in the expectation that their value will increase over time. Once the value has appreciated significantly, the investor sells the assets for a profit. The idea is to capitalize on market fluctuations. By buying when prices are low, investors aim to acquire assets at a discounted or undervalued price. When the market eventually recognizes the asset's true value and its price rises, the investor can sell it at a higher price than what they paid. Sounds promising, doesn't it? And simple, too! However, as I repeatedly explain to my children and my students, if it is too good to be true, then it is probably not true.

The approach is indeed too good to be true, and for several reasons. First, it is notoriously difficult to predict market movements accurately. As we learned previously, prediction and extrapolations are prone to significant errors. But, as we learned from the navy SEAL, we (may) win second by second. As time unfolds, more variables enter the system in a way that destroys all long-run predictions, except for trends available to all. Buying low and selling high relies on accurately identifying the optimal points to enter and exit the market, and timing the market consistently and accurately is a challenge. Even if you trust the predictions of wizard analysts and investment gurus, I advise you to check their success in making these predictions.

The second difficulty with the BLSH involves our emotions. We are emotional creatures, and those who lack emotions and try to think analytically without emotional involvement may perform worse. But emotional factors can also blur our judgment. For example, fear of missing out (FOMO) on an investment opportunity or fear of loss can lead to impulsive decisions, potentially causing investors to buy high and sell low instead.

Finally, sudden and unexpected market events can disrupt this strategy. Economic recessions, geopolitical events, or unforeseen crises can lead to rapid and unpredictable shifts in asset prices. And the long tail distribution has its own surprises. If you are a shoeshine boy who cannot abide with such things, it is better to avoid playing the game.

Despite all these difficulties, I will present one way of betting against the crowd by adopting the navy SEAL's approach. The basic idea is that the crowd's fallacy is to follow long-time trends and respond hysterically to catastrophic events or too enthusiastically to good news. Betting against the crowd, the individual should look for a niche the crowd is unaware of. In this chapter, the niche is the short time scale and the use of fluctuations that occur minute by minute.

There is nothing new about this approach. The famous Jim Simons, who launched a successful company, adopted the same approach. The first algorithmic trading companies made a similar creative move. However, I won't be presenting an algorithmic trading approach in this chapter. I will show an extremely simple heuristic that illustrates the importance of understanding and using constraints for short-term prediction and betting against the crowd by operating on a different time scale. Moreover, the time scale that I discuss is minute-by-minute. This time scale is significantly longer than the one used by algo-trading companies. Before presenting this, let me emphasize that this chapter presents no advice to the wise investor. It aims to illustrate the possibility of betting against the crowd by relying on foundational ideas and themes discussed in the previous chapters.

Betting Against the Crowd

In this section, I present the specific approach known as BLSH. The section is quite technical. However, the magic is in the details. First, let us represent a time series of stock prices as a series of permutations of length 3. We already discussed the representation of a time series by a series of permutations in Chap. 3, where we illustrated the importance of constraints for short-term prediction.

So, the first step is to obtain a time series of stock prices and convert the time series into a series of ordinal patterns: {0, 1, 2}, {0, 1, 2}, and so on. The first benefit from representing the stock prices by ordinal patterns is that, by focusing on order relations (e.g., the first stock price is higher than the second, which is higher than the third), we ignore the noise, an inherent part of any measurement process. We actually say that we are not interested in the absolute value of the stock price, but rather in the *relations* between three successive prices. Turning from absolute values to relations is a move that corresponds well to the way our minds work. The second benefit, as we learned before, is that constraints operate on the appearance of patterns and their transition. We can thus use them for prediction.

The approach used for the BLSH is simple. We focus our attention on a sliding window of three stock prices. This means that we look at three consecutive prices and then move one step to the right. So, effectively, we look at a very short segment of the data and move our window of attention one step to the right each time. We then repeat the process described here. The process is simple: whenever I observe a drop in the price from the first price to the second price (i.e., t_n to t_{n+1}), I buy $N \geq 1$ units of the stock and aim to sell it in the next step (i.e., the third step, or t_{n+2}) if an increase is observed (i.e., selling high). I move the window to the right and repeat the process: I buy when I see a decrease and sell if an increase follows. Therefore, the idea is simple. Do you see a drop in the price? If the answer is positive, then buy!

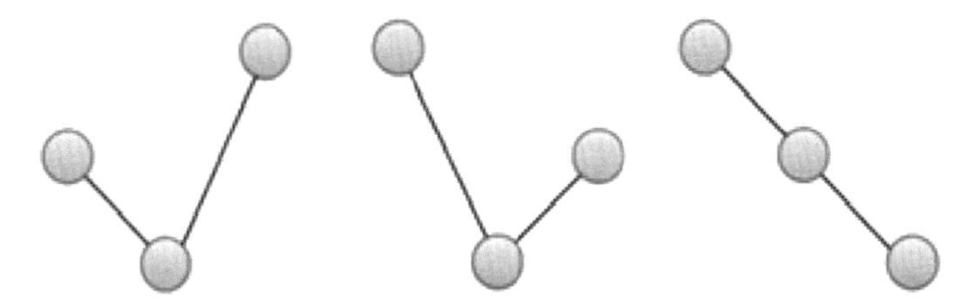

Fig. 5.1 A visual representation of the three permutations in which a drop in price is observed from t_n to t_{n+1}. *Source* Author

Now, there are three permutations where a drop in price is observed from t_n to t_{n+1}:

$$\pi_3 = \{1, 0, 2\}$$
$$\pi_5 = \{2, 0, 1\}$$
$$\pi_6 = \{2, 1, 0\}$$

These three permutations are presented visually, from left to right, in Fig. 5.1.

In each case shown in Fig. 5.1, there is a fall in price from the first price to the second. We may profit from selling the stock at the third step in two of the above cases. In one case (i.e., π_6), a loss is experienced when selling the stock in the third step. In two cases, permutations 3 and 5, the BLSH is a successful strategy, whereas in the third case, permutation 6, it is a failure. So, to repeat, we observe two consecutive stock prices and if the second price is lower than the first, we buy the stock. Then, if the price increases in the third step, we can sell it and make a profit. *This strategy holds for only two of the permutations.* However, if we accidentally fall on permutation 6 (i.e., 2, 1, 0), then selling the stock at the third step ends in a loss, because the third stock price is lower than the second.

In sum, if we examine the minute-by-minute prices of a stock and can immediately respond by buying or selling, then in two out of three permutations (i.e., ordinal patterns) involving a drop in price in the second step, I can gain by selling the stock in the third step and re-entering the game. Sounds promising, as long as the probability of encountering a monotonic decreasing pattern (i.e., 2, 1, 0) is low enough.

At this point, we may understand that the BLSH approach I presented is built on the *assumption* that there is a higher probability of observing permutations 3 and 5 in the time series than the monotonic decreasing permutation, permutation 6 (i.e., 2, 1, 0). Therefore, the minimal entrance point (MEP) to our BLSH game, as described above, is

$$\mathrm{MEP} = 100 * \frac{\mathrm{n}(\pi 3 + \pi 5)}{\mathrm{n}(\pi 3 + \pi 5 + \pi 6)} \geq 51$$

meaning that the relative frequency of permutations 3 and 5 in the time series of permutations, expressed as a percentage, is greater than or equal to 51. The MEP is a simple heuristic for selecting stocks for trading and applying the simple BLSH strategy described above. For example, together with Yochai Cohen, I analyzed the time series of 500 stocks[1] (2017-09-11 to 2018-02-16), represented the time series as a series of permutations, and for each stock computed its MEP and the 95% confidence interval (CI) for the MEP. The data covers a time range of six months.

The 95% confidence interval is a statistical way to determine the probable boundaries of our measure. For example, the MEP of the stock known as NWS is 87.80. By analyzing the time series of this stock's prices, we found that ordinal patterns 3 and 5 significantly dominate. However, 87.80 is a measurement taken from a sample that does not necessarily represent the population of data points. The confidence interval tells us that under certain statistical assumptions, the true value of the MEP probably falls somewhere between 86.78 and 88.83.

Sorting the stocks according to the *lower bound* of their MEP 95% confidence interval, I identified and selected the top seven stocks and used them for the experiments (see Table 5.1).

We see that all of the selected stocks are ranked high on our minimal entrance point to the game. The percentage of the monotonic increasing pattern among these stocks is relatively high. However, this is only the minimal entrance point to the game, and successfully playing the game requires much more. According to the above-mentioned idea, selecting stocks

Table 5.1 The top seven stocks, according to the lower bound of the MEP

STOCK	MEP	SD	CI_LOW
NWS	87.80	5.46	86.78
AES	76.01	6.40	74.80
NWSA	74.54	6.97	73.23
CHK	73.71	5.08	72.75
F	72.14	6.22	70.97
HPE	71.56	7.43	70.16
WU	70.90	6.73	69.63

[1] https://www.kaggle.com/nickdl/snp-500-intraday-data.

according to their MEP should result in a beneficial BLSH strategy. This is, of course, as long as the value of the MEP holds out. Reality may change, and so may the MEP. Therefore, in practice, we should carefully monitor the MEP and its fluctuations. Moreover, the MEP is a necessary, but insufficient condition for a successful BLSH strategy. The next step is to make a quick estimate of the *expected value* of applying the BLSH.

What Is Your Expected Value?

The idea of the expected value is very important, and is highly recommended by the legendary investor Warren Buffett:

> Take the probability of loss times the amount of possible loss from the probability of gain times the amount of possible gain. That is what we're trying to do. It's imperfect, but that's what it's all about. Warren Buffett, quoted in Griffin (2015)

To compute the expected value, we must somehow estimate the outcome of selling the stock at the third component of each permutation in two scenarios: gain and loss. For simplicity, we assume that, in the context of high-frequency minute-by-minute trading, the average difference in price from minute to minute, denoted by Δ, is relatively stable and small. Table 5.2 presents the average Δ of the above-mentioned stocks in absolute percentages.

We can see that the average minute-by-minute difference in prices is small. Therefore, for permutation 5 (i.e., 2, 0, 1), where a price drop is observed, followed by an increase, we may heuristically assume that the price increase is half the value given by Δ. Given our ignorance of the size of the drop from rank 2 to rank 0, it is a simplification to assume that the increase in price from rank 0 to rank 1 is half the size of the value given by Δ. In this case, buying a stock at the price ω and selling at $\omega + 0.5\Delta$ would leave us with

Table 5.2 The average delta in the stock price

STOCK	% Δ
NWS	0.0001
AES	0.00002
NWSA	0.00008
CHK	0.00002
F	0.00002
HPE	0.00007
WU	0.00003

a gain of 0.5Δ. According to the same logic, for permutation 3, the average expected gain is 1.5Δ. In all cases where we buy a stock after observing a price drop, the loss expected after observing another drop (permutation 6) is Δ. By a simple analysis:

$$\text{Expected Value} = \sum P(Xi)^{*}Xi$$

We should expect that in adopting a simple BLSH strategy, where we gain whenever we encounter permutations 3 and 5 and lose whenever we encounter permutation 6, the expected value will be *positive*, meaning that applying the BLSH strategy to the series of permutations under the constraint imposed by the MEP will be a beneficial strategy. In other words, if the MEP is high enough and the difference in the stock price is small and consistent, then *repeatedly* applying the BLSH strategy to the stocks selected using the MEP heuristics should (in all probability) be beneficial.

So, this is another step, but still not enough. In the previous chapters, I have emphasized the importance of being non-stupid. One of the ways of addressing this challenge, at least in the limited context of repeating games of chance, is the Kelly criterion.

The Kelly Criterion

Now that we have a "winning" strategy, we may bet against the crowd by choosing a short time window and using the constraints imposed by the permutations. The constraints are that, in 2 out of 3 cases of falling prices, we should observe a third value higher than the second. The MEP assures us that we trade stocks where this constraint is probabilistically established. However, "probabilistically established" sounds rather like an oxymoron. The MEP may change, and we must monitor it carefully. Moreover, risking our money in a probabilistic game requires some caution, and the Kelly criterion will be the lighthouse that serves to safely bring us to the shore.

What is the Kelly criterion? When learning about the Kelly criterion, understand that it may be one of the most brilliant ideas you have ever learned. The Kelly criterion deals with a simple question: How much of your wealth should you invest in a repeated bet or investment? Suppose you are a shoeshine boy eager to participate in the stock market rush. You have a total wealth of 5$. How much of it can you risk on the stock market? Now, ask the same question for Joe Kennedy. Wikipedia tells us[2] that, in 1929, Kennedy's fortune was estimated to be $4 million. Assuming that only some of this was available for investment, we can ask the same question for Joe. Given a bankroll of $1,000,000 available for investment, how much was it rational to invest in the market?

The question sounds extremely difficult to answer. However, Kelly proposed that a bet's optimal size depends on the probability of success and the amount we stand to gain or lose. Let me give you a simple example. You are visiting Las Vegas and are seduced to enter one of the casinos, where you meet the one-armed bandit. The one-armed bandit gained a certain reputation because the probability of winning on any given spin may be extremely low. We assume that, when you spin the wheel, there is a chance of obtaining a winning combination in 1% of the cases. Well, that doesn't sound very promising, but this is why this gambling machine is called a bandit. However, the atmosphere, the food, and the alcohol limit the activity of your frontal cortex, and you decide to bet.

Suddenly, Kelly pops into your mind, advising you to think about the amount you are ready to risk. But with optimal timing, a beautiful lady then appears with another free chaser. "Need any help?" she asks. You say, "I'm just thinking about how much money I should invest. My total bankroll is only $10,000, as I am a retired shoeshine boy." The lady, who was sent by the bandit's big boss, smiles and offers you another free chaser. "Don't worry," she says, "for each $1 you invest in each spin, you may gain a lot of money if you win." "How much?" you ask. "A lot!" She replies and offers you another free chaser. "I offer you odds of 10,000:1. For every dollar you bet, you could win $10,000 if the bet is successful." This sounds like an amazing offer. You risk only $1 in each spin but you gain $10,000 if you win! Your amygdala bursts with excitement.

Fortunately, at this point, your cortex awakens and interferes in the conversation, before you put your hand in your pocket. It suggests using the Kelly criterion:

$$f^* = \frac{bp - q}{b}$$

where f^* is the fraction of the current bankroll you should risk, p is the probability of success, q is the probability of failure, and b is the odds, or the ratio of the amount you stand to win to the amount you stand to lose. Based on the Kelly criterion, your cortex advises you to risk only 1% of your

[2] https://en.wikipedia.org/wiki/Joseph_P._Kennedy_Sr.#cite_note-inflation-US-12.

bankroll; otherwise, you are doomed to bankruptcy after repeatedly playing the game. Not even one cent more unless you are stupid and wish to end your night in bankruptcy.

Moreover, your cortex is even more cautious. It advises you to be much more conservative and to play with a *fraction* of the proposed betting size, for example, with only 0.5% of your wealth. If you are a retired shoeshine boy, do not play with your new bandit friend for over $50. Beyond the Kelly criterion, there is a dangerous betting area: *hic sunt dracones*.

The Kelly criterion is a wonderful idea if you are playing a repeated game and you know the probability of winning and the odds. But is it applicable to our idea of buying low and selling high? Let us see.

Playing with Kelly

In our case, b can be estimated by simulating the time series, but we may also assume that since, in the short run of a one-minute difference, the profits from winning are the same as those from losing. Given this assumption, we may set b equal to "1". If the opening price of a stock is 11.23 USD and p = 63% (rounded), the Kelly score is 50%. However, as Kelly represents the *limit* of a rational bet, a more cautious strategy would be to use a fraction of Kelly, such as 0.33 of the value of f^*. The proportional Kelly (0.33) score is 17%. In this case, our optimal betting size would be up to 17% of our total bankroll. This limit is important to the rational agent because, in the above context, it provides her with a simple heuristic for deciding whether it is beneficial to enter the BLSH procedure described above. For example, if the price of a single stock is 11.23 USD and Kelly advises you to risk up to 17% of your total bankroll (BR), then your total bankroll or your "pocket size" must be at least

$$BR = \frac{100 * Price}{f^*} = \frac{100 * 11.23}{17} = 66 \text{ USD}$$

for trading with a single stock at a time. This means that, unless you have a pocket size of $66, it is better not to enter the game given the above constraints. The overall heuristic presented so far may be generally described as follows:

1. Search for stocks satisfying the MEP criterion.
2. Rank the stocks according to their MEP score and select the top k stocks.
3. For each stock, compute the proportional Kelly score.

4. Use the price of a single unit to determine the minimal "pocket size" (i.e., BR).
5. If the minimal pocket size is higher than your actual pocket size, refrain from playing the game; otherwise, apply the BLSH strategy.
6. When using the BLSH strategy, buy whenever you observe a fall in price and sell immediately afterward.

To test the above heuristic, I analyzed the minute-by-minute stock prices of the seven stocks presented above. I tested three versions of the heuristics: fight, flight, and freeze. The procedure is detailed in the next section. For the first experiment, I used a simple heuristic. For a given selected stock:

1. Define a bankroll (BR), the sum you begin with.
2. Start with the first minute of your time series.
3. Your focal point is always defined as t_n the minute you observe.
4. If you observe a decrease in price from t_{n-1} to t_n, buy ONE STOCK.
5. If you observe an increase from t_n to t_{n+1}, THEN sell, update your BR, and restart the game from t_{n+2}. At this point, t_{n+2} is your focal point.

I experimented with three versions of the heuristic. The different strategies differ concerning the situation where you observe a *decrease* from t_n to t_{n+1} (when you actually observe permutation 6). The first strategy is *freeze*: whenever you observe another price drop, wait for a price HIGHER than the one observed on your "buy point" and only then sell. The second strategy is *flight*: whenever you observe another price drop, sell the stock at t_{n+1}, update the BR, and restart the game with your focal point now at t_{n+2}.

The third and the most complicated strategy is *fight*, a rational Martingale strategy. The *Martingale* is a well-known class of failed betting strategies where the gambler doubles his bets upon losses. It is a failed strategy as the exponential growth of the bets leads to bankruptcy. Here, I propose a rational version of this strategy. Whenever a price decline is observed from the second to the third value of the permutation (i.e., permutation 6), use the proportional Kelly score of your updated BR and compute the possible length of a Martingale buying sequence you can risk without experiencing bankruptcy (e.g., 2 stocks, 4 stocks, 8 stocks). Now start buying according to the Martingale principle. For instance, if you have money, you can buy two stocks at t_{n+1}, 4 at t_{n+2}, etc. You sell when the price increases or when you run out of money, GIVEN the sum you originally dedicated to the Martingale, and then restart the game.

It is important to notice that entering this form of rational Martingale is justified only if:

$$P(\pi 6 \rightarrow \pi 6) < 0.5$$

which means that the probability of a transition from permutation 6 to permutation 6 is lower than 0.5. For example, if we find that

$$P(\pi 6 \rightarrow \pi 6) = 0.355$$

then, assuming independence from one step to the next, we may predict that the probability of a second transition to $\pi 6$. is 0.126, a third transition to $\pi 6$ is 0.045, a fourth transition is 0.016, and so on. This information allows us to approximate the possible length of a series of continuous price falls (i.e., a series of instances of permutation 6) and the expected length of Martingale sequence you may be required to apply.

So, Have We Won?

Table 5.3 presents the price of a single stock for the first point of the time series (FIRST), for the last point of the time series (LAST), and the maximum price observed through the period (MAX). Figure 5.2 presents the final profit for the three strategies applied to the AES time-series.

We can see that the flight strategy outperformed the others. Table 5.4 presents the gain for the flight strategy, with an average gain of +2.23%. This means that, across all the stocks that we have analyzed, we completed our experiment with a positive but very modest profit of approximately 2% for trading one stock at a time.

Table 5.3 First, last, and maximum price of the stock during the tested period

Stock	First	Last	Max
NWS	13.45	16.40	17.70
AES	11.24	10.47	11.95
NWSA	13.23	16.28	17.28
CHK	3.66	2.72	4.49
F	11.40	10.64	13.47
HPE	13.25	16.40	17.06
WU	18.79	20.27	22.17

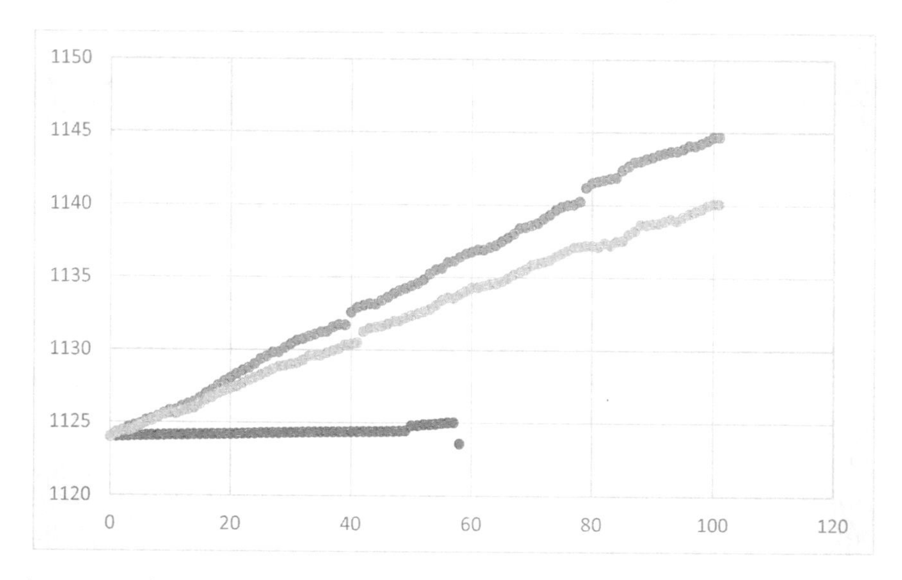

Fig. 5.2 Profit as a function of time for the three versions of the BLSH strategy. Freeze is represented by the lowest line and flight by the uppermost line. *Source* Author

Table 5.4 Percentage gain for the flight strategy

Stock	Initial BR	Gain	% Gain
NWS	1345	+ 83.10	6.2
AES	1124	+ 21.00	1.9
NWSA	1323	+ 17.55	1.3
CHK	336	+ 17.58	4.8
F	1140	+ 19.7	1.7
HPE	1325	+ 17.44	1.3
WU	1879	+ 17.33	0.9

The above results support the hypothesis that a permutation-based representation and the MEP heuristic applied to this representation can be at least modestly profitable. However, in the first experiment, I bought or sold only *one stock unit* at a time. What would have happened if, instead of buying a single stock, we had used the relative Kelly score of our BR to buy N stocks?

Experiment 2 replicated the above procedure with a minor but extremely important variation: I bought N stocks at a time instead of buying a single stock. I started the trading procedure using a BR equivalent to 100 times the

Table 5.5 Percentage gain for buying N stocks using the flight strategy

Stock	Initial BR	Gain in $	% Gain
NWS	1345	+ 4778	355
AES	1124	+ 616	54.8
NWSA	1323	+ 391.21	29.56
CHK	336	+ 627.92	172
F	1140	+ 467.95	41
HPE	1325	+ 385.58	29.1
WU	1879	+ 359.78	27

initial price of the stock. The initial BR was $1345 which is equivalent to the value of 100 stocks. Table 5.5 presents the gain for each stock in the flight strategy.

When trading N stocks, we see that the BLSH flight strategy resulted in an average gain of 101.20%. This means that the proposed heuristics gained an average of close to 100% profit at the end of the trading period! As you can see, betting against the crowd is possible.

Simple Models in a Complex World

Buy low, sell high is one of the basic rules of thumb individuals use for investment (Lamas, 2021). However, the apparent simplicity of this heuristic is repeatedly questioned. As argued on a professional investment site,[3] "It's so obvious it sounds like a joke. In reality, it is a lot easier said than done" because predicting the optimal timing for buying/selling the stock involves a "crystal ball" (Makridakis & Taleb, 2009) that just doesn't exist for such situations. While it seems obvious to use prediction/forecasting to apply the BLSH strategy, we should recall the humorous advice from Makridakis et al., (2010, p. 88): "By all means, make forecasts – just don't believe them". What is hard to belief is that any pointwise predictive model would allow us to buy at the lowest price and sell at the highest.

In this chapter, I have shown that representing a time series of stock prices as a series of permutations may be the first step in a successful heuristic BLSH trading strategy. Using a permutation-based representation has several benefits, such as moving from absolute to relative values and de-noising the series. Moreover, simply calculating the MEP gives us the first indication of which

[3] https://www.investopedia.com/articles/investing/081415/look-buy-low-sell-high-strategy.asp.

stocks we should select for trading, and by applying the Kelly criterion, we may successfully trade without the risk of bankruptcy.

This approach favoring simplicity avoids the pitfall of prediction that is usually associated with BLSH. Mousavi and Gigerenzer (2017) explain that heuristics are highly important tools for uncertainty in situations where the edge, or the competitive advantage, lies in the uncertainty. Therefore, instead of reducing uncertainty, the appropriate trading heuristic may actually exploit it. The uncertainty associated with fluctuations in stock prices may be used by our proposed heuristic, which avoids the pitfalls of prediction, and the benefit of its simplicity is apparent. Explaining this benefit, we may recall Katsikopoulos et al., (2018, p. 20), who argued that "for complex models to outperform simple ones, two conditions must be met: (1) The real-world process must be complex, and (2) the forecaster must be able to model this complexity correctly". High-frequency trading is complex, and we cannot accurately model this complexity. Avoiding prediction and relying on a simple model and heuristic may therefore be a reasonable strategy.

Lessons to Be Learned About Betting Against the Crowd

What lesson can we learn for understanding crowds and betting against them? The first lesson is that betting against the crowd is a risky business. One should give it careful thought before deciding to swim against the current. The second lesson is that betting against the crowd is possible if you understand the crowd's dynamics and the constraints operating on this dynamics. The third lesson is that identifying the relevant time scale for gaining an edge is critical for success. I have repeated this point and will keep repeating it. Identifying the relevant time scale for intervention is crucial, and this time scale for intervention must be different from the one in which the crowd's behavior evolves. Although we may think of time as equal for all, we live in a world where different watches tick. Hitting a potentially violent aggressor a fracture of a second before he moves from intention to action may be a difference that makes a difference. Similarly, deceiving yourself that you are an eternal being may be a sweet illusion but it is one with detrimental consequences. Sometimes, we should think of ourselves as short-lived creatures like bees. Sweet consequences may follow by working on the time scale of a bee or a butterfly.

The final lesson is that, even if you are creative enough to bet against the crowd repeatedly, you must be cautious, as proposed by Kelly. The Kelly criterion is relevant for a very specific type of repeating game. In life, you may encounter contexts where you have only one shot. Understanding which game you are playing is very important. Pretending to play a repeating game when you actually have only one shot is a recipe for disaster.

In sum, I used the example of buy-low-sell-high to illustrate the possibility of betting against the crowd in a specific context, which is the stock market. I pointed to the importance of (1) identifying a time frame where you may have the edge over the crowd, (2) using short-term prediction instead of long-term beliefs and fantasies, (3) using a simple but profound algorithm to decide how to play, and (4) taking some cautious measures to limit your risk and steer clear of the area where dragons roam.

References

Griffin, T. J. (2015). *Charlie Munger: The complete investor*. Columbia University Press.

Katsikopoulos, K. V., Durbach, I. N., & Stewart, T. J. (2018). When should we use simple decision models? A synthesis of various research strands. *Omega, 81*, 17–25.

Lamas, S. (2021). Simple decision-making rules of thumb and financial well-being. *Journal of Personal Finance, 20*(2), 22–39.

Makridakis, S., Hogarth, R. M., & Gaba, A. (2010). Why forecasts fail. What to do instead. *MIT Sloan Management Review, 51*(2), 83.

Makridakis, S., & Taleb, N. (2009). Living in a world of low levels of predictability. *International Journal of Forecasting, 25*(4), 840–844.

Mousavi, S., & Gigerenzer, G. (2017). Heuristics are tools for uncertainty. *Homo Oeconomicus, 34*(4), 361–379.

Neuman, Y., Cohen, Y., & Tamir, B. (2021). Short-term prediction through ordinal patterns. *Royal Society Open Science, 8*(1), 201011.

Part II

Case Studies: Political Nationalism, Football, Financial Markets and Armed Conflicts

6

Non-linearity in the Emergence of Political Nationalism: A Lesson from Hungary

Introduction

When discussing crowds, one immediate association comes from recollections of our near past. Educated people must have watched documentary films where Hitler incited his German mob. A book dealing with human collectives cannot ignore politics, as we are still quite ignorant about political crowds, despite numerous books and papers written on the subject. Again, these crowds are both simple and complex. Listening to commentators, journalists, and "experts" in political science, we might easily conclude that we have not gained significant insights beyond those already provided by lay observations. On the one hand, the behavior of crowds remains obvious after listening to these commentators, while on the other hand, what is non-obvious in crowd behavior is still non-obvious.

In this chapter, I aim to provide a fresh perspective, at least with respect to certain kinds of political crowd behavior. As mentioned above, a crowd differs from the sum of its parts. Therefore, the behavior of the crowd involves an important component of uncertainty. Moreover, crowds are not naturally homogenous. They include some variability, which may be important for understanding their behavior. I will show how this variability may play a role in the formation of the whole.

The subject of the current chapter is the rise of nationalism and right-wing extremism in Europe. This phenomenon is of great concern to those who still remember the cost of nationalist authoritarian regimes. My family, who survived the holocaust, gave me a first-person testimony of the atrocities perpetrated by Hitler and his collaborators. For me, right-wing nationalism is

Y. Neuman, *Betting Against the Crowd*, https://doi.org/10.1007/978-3-031-52019-8_6

a personal issue, like its mirror image, left-wing Stalinism, or any other form of zealous *dogma* seeking to impose itself on others regardless of common sense, humanity, and its often disastrous consequences.

In Europe, there are now several nationalistic governments. For example, Poland is currently governed by the right-wing nationalists of the Law and Justice Party, which is portrayed as anti-Western (Müller, 2021), silencing internal criticism and masking Poland's dark history (Gessen, 2021). Similarly, under the leadership of Victor Orban and Fidesz, Hungary is accused of playing a 'double game with antisemitism' (Kingsley, 2019), a classic sign of right-wing nationalism. It also follows an anti-critical policy by silencing liberal forces, for instance, by forcing out the European university sponsored by George Soros (Thorpe, 2020).

In these cases and others, we may ask whether these expressions of nationalism should be naturally expected from the simple accommodation of nationalist forces comprising the government or whether they are the *nonlinear* result of more complex interactions. In other words, the question is whether, and to what extent, nationalism is the simple additive outcome of political forces operating in the political system or the outcome of nonlinear dynamics, which is more complex and surprising (West, 2016).

The answer is probably a mixture of the two. Sometimes, a strong and homogenous political force may be the sole explanation. Hitler's regime is a classic case. However, in such a case, no research is required. What you see is what you will get. The situation may be more complex when the political system involves some variability. In this context, the question of how several political units interact to form a whole that differs from the sum of its parts may be of great interest. This question is of interest even for political systems outside Europe. For example, Israel has recently been experiencing a struggle between a coalition of forces threatening Israel's democracy and a liberal civilian protest. This threat to Israeli democracy has come as a surprise to most Israelis, like a freak wave unexpectedly rocking a ship on the ocean.

In this context, we may ask whether the surprise experienced by the Israelis results from their blindness to the accommodating forces or from the nonlinear dynamics of their current government. The analogy between freak waves and unexpected changes in a political system may be highly constructive. Freak waves are huge ocean waves that violate our "normal" expectations. These unexpected waves can be extremely dangerous for ships because they are characterized by their suddenness, immense height, and steepness. Rogue waves can reach over twice the average of the tallest waves around them. They often appear seemingly out of nowhere, which makes them particularly hazardous for ships at sea.

The exact cause of rogue waves is not always understood. However, they can result from various factors, including the interaction of different wave patterns, ocean currents, and wind conditions. In the past, rogue or freak waves were considered to be delusions imagined by drunk sailors, simply because they were considered so unlikely under the assumption of the Gaussian (i.e., Normal) distribution (DW Documentary, Undated). However, the scientific evidence for the existence of rogue waves has been established beyond doubt using new measurement technologies (Dysthe et al., 2008).

The mechanisms producing such waves are still debated, but one appealing explanation suggests that they may be formed through *non-linear interactions* of waves comprising a whole *bigger than the sum of its parts*. Here, we immediately understand the similarity between freak waves and governments. Currently, we have no rigorous scientific methodology for measuring the non-linear interactions expressed in a government's level of nationalism. In this chapter, I propose an approach and illustrate its use by analyzing Hungarian governments since 1990 (Wikipedia A, Undated).

Measuring the Expected and the Observed Levels of Nationalism

My analysis focuses on the Hungarian government and, more specifically, on the historical composition of the National Assembly of Hungary since 1990 (Wikipedia B, Undated). For the analysis, I used the data that appears in Wikipedia. To determine the *expected degree* of nationalism in a government, I used the political position of the parties comprising the government. Wikipedia labeled the party's political position from center left to right-wing and far right. Therefore, I scored the political position of each Hungarian party on a five-point ordinal scale ranging from center-left (score = 1) to far right (score = 5).

I first computed the expected nationalism score of a government using three measures. The first is the expected political position of the government, based on a simple *average* of the position scores of the parties composing it. This measure is called expAVG. For example, the first Orban government comprised three parties: Fidesz (score 5), FKgP (score 3), and MDF (score 3). The average expected nationalism score is therefore "3.67". I have also calculated the *weighted* score for the government based on the percentage of seats held by the government parties in the national assembly (Wikipedia B, Undated). For example, between 1998 and 2000, three parties held 148, 48, and 17 seats for the first Orban government. I have converted these numbers to percentages and multiplied each percentage by the nationalism score of the party:

$$\text{Expected nationalism score} = (0.69 * 5) + (0.23 * 4) + (0.08 * 3) = 4.61$$

Based on this weighted average, the first Orban government was expected to be far to the right of what would be expected from a simple average. This measure is called expWEI. The third and final score was the geometric mean of the nationalist score of the parties (i.e., expGEO). The geometric mean is often used for "a set of numbers whose values are meant to be multiplied together or are exponential in nature." (Wikipedia). I used this as a third measure.

So far, I have calculated three expected or predicted scores based on the parties' political positions. The next step is to evaluate the *government's actual/observed political position.* The natural way to determine the nationalism score of a government is to turn to experts in political science. However, as the judgment of human expertise in the social sciences might be severely biased, and anyway my trust in academic "experts" from the social sciences is limited, I took an unusual approach: I used AI to measure the actual nationalist score of each government. More specifically, I used the most sophisticated language model—GPT.

GPT is an extremely sophisticated language model. This means that it has learned contextual language patterns from a huge data repository. This language model can leverage almost all aspects of social and psychological research, as I have shown elsewhere (Neuman & Cohen, 2023). The trick is, of course, to use the tool wisely and modestly. Naturally, chatGPT is not omniscient, and asking it questions about the existence of God or whether you should propose marriage to your girlfriend is beyond its capacity. On the other hand, it is an extremely intelligent device that can answer various questions and perform other intelligence tasks. But, to get reasonable answers from chatGPT, one must ask the right kind of questions and present them in a format that the machine can use effectively. This format is called a "prompt," and designing effective prompts is an art no less than a science.

To generate the nationalism score of each government, I used the following prompt by first introducing the machine to the context relevant to its task:

Context: Here are some common criteria for evaluating the political position of a government:

1. **Ideological Beliefs and Values**:

- **Nationalism**: This can be measured by examining the government's commitment to promoting and protecting its nation's interests, culture, and identity. A strong emphasis on national pride and loyalty, and a preference for domestic policies over international cooperation can indicate a nationalist stance.
- **Right-Wing**: Right-wing ideologies often prioritize individualism, limited government intervention, free markets, traditional values, and a strong emphasis on law and order.

2. **Economic Policies**:
   - **Nationalism:** Economic policies prioritizing domestic industries, protectionist trade measures, and economic self-sufficiency can indicate nationalist tendencies.
   - **Right-Wing**: Right-wing economic policies typically favor minimal government intervention in the economy, lower taxes, deregulation, and free-market capitalism.
3. **Social and Cultural Views**:
   - **Nationalism:** Strong attachment to the cultural and social norms of the nation, a focus on preserving traditional values, and resistance to cultural assimilation or globalization can suggest nationalist sentiments.
   - **Right-Wing**: Right-wing individuals and groups often uphold traditional social norms, family values, and religious principles. They may also advocate for stricter immigration policies.
4. **Attitudes Toward Immigration**:
   - **Nationalism**: Strong opposition to immigration can indicate nationalist inclinations, especially if framed as protecting the nation's identity and resources.
   - **Right-Wing**: Right-wing ideologies might generally advocate for stricter immigration policies, emphasizing national security and cultural cohesion.
5. **Foreign Policy and International Relations**:
   - **Nationalism**: Focusing on pursuing policies that prioritize the nation's interests above international cooperation or treaties can reflect nationalist viewpoints.
   - **Right-Wing**: Right-wing foreign policy might lean towards a more assertive stance, emphasizing military strength and protecting national sovereignty.

6. **Attitudes Toward Social Welfare**:
   - **Nationalism**: Advocacy for social welfare policies that benefit the nation's citizens over outsiders can suggest nationalist perspectives.
   - **Right-Wing**: Right-wing ideologies often emphasize personal responsibility and limited government involvement in social welfare programs.
7. **Use of Symbols and Rhetoric**:
   - **Nationalism**: Frequent use of national symbols, slogans, and appeals to national unity can signal a nationalist orientation.
   - **Right-Wing**: Law-and-order rhetoric, appeals to traditional values, and strong support for patriotism are often associated with right-wing ideologies.
8. **Political Alliances and Affiliations**:
   - **Nationalism**: Collaboration with other nationalist or ethnocentric groups can provide evidence of a nationalist leaning.
   - **Right-Wing**: Associations with conservative political parties, right-wing think tanks, or advocacy groups can indicate a right-wing orientation.

This context allows the language model to direct attention to the relevant parts of its digital brain. After presenting the context, I defined a specific task to perform:

Task: Score each Hungarian government on a five-point scale. The scale indicates the government's nationalism. The scores are: "1" (center left), "2" (center), "3" (center right), "4" (right wing), and "5" (right-wing and far right). Produce a single score for each of the following Hungarian governments:

1. Antall
2. Boross
3. Horn
4. Orbán I (July 6, 1998–May 27, 2002)
5. Medgyessy
6. Gyurcsány I (September 29, 2004–June 9, 2006)
7. Gyurcsány II (June 9, 2006–April 14, 2009)
8. Bajnai (April 14, 2009–May 29, 2010)
9. Orbán II (May 29, 2010–June 6, 2014)
10. Orbán III (June 6, 2014–May 18, 2018)
11. Orbán IV (May 18, 2018–May 24, 2022)

12. Orbán V (May 24, 2022–present).

Given the above-mentioned task, ChatGPT generated the following scores:

1. Antall: Center-Right (3)
2. Boross: Center-Right (3)
3. Horn: Center-Left (1)
4. Orbán I: Center-Right (3)
5. Medgyessy: Center-Left (1)
6. Gyurcsány I: Center-Left (1)
7. Gyurcsány II: Center-Left (1)
8. Bajnai: Center-Left (1)
9. Orbán II: Right-Wing (4)
10. Orbán III: Right-Wing (4)
11. Orbán IV: Right-Wing (4)
12. Orbán V: Right-Wing (4.5).

I used these scores to compute the nationalism scores of the nine political National Assemblies forming governments in Hungary since 1990. More specifically, I used the "Historical composition of the National Assembly" since 1990 (Wikipedia B, Undated). The nine compositions are numbered GOV 1 to 9. For example, the composition of 1990–1994 overlaps with Antall's government; GOV 3 corresponds to the first Orbán government, and compositions 7–9 correspond to the last Orbán governments.

Next, I measured the Pearson correlation between the nationalism score produced by AI and the three expected nationalism scores I calculated by averaging the nationalism scores of the parties according to Wikipedia. I hypothesized that, if the AI had given a valid measurement of the nationalism score, a positive and significant correlation should be found between the AI score and the three expected scores I had calculated.

Given the small size of my observations ($N = 9$), and in addition to the correlation, I also calculated two additional statistical measures. The first measure is the Bayes factor (BF). This quantifies the strength of evidence in favor of one hypothesis compared to another. In our case, the hypothesis was that the score computed by AI was positively correlated with the expected scores. The BF is "1" if the evidence is balanced and greater than one if the evidence favors the hypothesis. In addition, I calculated the Vovk–Sellke maximum p-ratio (VS-MPR), which measures the maximal possible odds in favor of our hypothesis over the alternative hypothesis. The higher the scores, the better supported the hypothesis. The results are presented in Table 6.1.

Table 6.1 Correlation between the AI nationalism score and the three expected scores

Expected	r	BF	VS-MPR
ExpAVG	0.98	9060	34,619
ExpWEI	0.93	217	327
ExpGEO	0.96	1063	2291

The correlation results indicate that the score proposed by AI is highly and significantly correlated with the scores expected by averaging the nationalism scores of the parties comprising the government. The most significant correlation is between the score computed by AI and the *average expected score*. This result is interesting, as it suggests that the AI system may have been biased in a similarly way to our naïve human understanding, expecting the whole to be some average of its constitutive parts. In a deep sense, AI seems deeply flawed when it comes to understanding complex social systems as it averages the components to understand the whole. The interesting part, however, is that in some cases, and regardless of the high correlation, there is a *gap* between the expected and the observed nationalism score, as you can see in Table 6.2.

For example, we can see that the first Orbán government (GOV 3) scored significantly *lower* than expected. It means that it was less nationalist than expected. How significant is this gap between the expected and the observed score? To answer this question, we can compute the median absolute deviation (MAD) for the gap between the expected and observed scores.

The median is the middle value in a data set when the data is arranged in ascending or descending order. In other words, it is the value that separates the upper half from the lower half of a data set. MAD is a measure of variability defined as the median of the absolute deviations from the median of

Table 6.2 The gap between the expected and observed nationalism scores. The observed score is the one generated by GPT

GOV	EXP	OBS	GAP
1	2.08	3	−0.92
2	1.25	1	0.25
3	4.61	3	1.61
4	1.10	1	0.1
5	1.10	1	0.1
6	4.86	4	0.86
7	4.88	4	0.88
8	4.88	4	0.88
9	4.87	4.5	0.37

the data. It means that we take a data set and first compute the deviation (or distance) of each observation from the median. We look at the absolute deviations, meaning that we ignore whether a deviation is positive or negative, and compute the "middle value" of these deviations. This is MAD. For the gap between the expected and the observed score in my data set, the median is 0.37 and MAD = 0.49.

Next, I calculated the *absolute* difference between each data point (i.e., the difference between expected and observed) and the median in terms of MAD. This procedure allows us to understand how far an observation is from the median using MAD dispersion units. For example, the result for the first Orbán government is

$$\frac{|Xi - Median|}{MAD} = \frac{|1.61 - 0.37|}{0.49} = 2.53$$

This means that the difference between the expected nationalism score and the actual nationalism score of the first Orbán government, as measured by AI, is 2.53 dispersion units from the median. This number suggests a significant gap or error in predicting the nationalism level of the first Orbán government, and the same is true for GOV 1, led by Antall.

My interest lies in explaining this gap using simple scientific measures. To estimate the extent to which nonlinear interactions may be responsible for the gap between the expected and observed scores, I measured the entropy of each of the nine assemblies using the distribution of the parties composing the government. More specifically, I analyzed the distribution of forces within each government. For instance, the first Antall government was composed of the following parties: MDF (164 seats) and SZDSZ (93 seats). Turning this distribution into percentages, we get 64% and 36%, respectively. However, the important measure of variability is discussed in the next section.

Entropy, Variability, and Non-linearity

Previously, I explained that Shannon information entropy is best described as a measure of variability indicating the extent to which a certain distribution is more or less variable. For instance, the first Orbán government comprised three political parties: Fidesz, FKGP, and MDF. The Shannon entropy of this government is higher than the entropy/variability of the current Orbán government, made up of Fidesz and FKGP alone. The variability may be highly important for understanding social systems (Neuman, 2021) because it may indicate how the variability of political forces is associated with the *overall political position* of the government. To measure the variability of political composition, I did not use the Shannon entropy, but rather the Tsallis entropy (Tsallis, 2014), an entropy measure that can be used for complex social systems. To explain this measure, we need to discuss the ideas of *additivity* and *extensivity*.

Additivity and Extensivity

In physics, additivity and extensivity describe the behavior of physical quantities when combining or scaling systems. While they are related, they have distinct meanings. Additivity refers to the property of a physical quantity that allows it to be *combined* by simple addition when multiple systems are present. In other words, if a quantity is additive, the total value of that quantity for a composite system is equal to the sum of the values for the individual components. For example, if you have two identical containers of gas, the total volume of the combined system is simply the sum of the volumes of the individual containers. Similarly, the total energy of a system consisting of two non-interacting subsystems is the sum of the energies of the subsystems. Extensivity, on the other hand, refers to the property of a physical quantity that scales linearly with the size or extent of a system. If a quantity is extensive, doubling a system's size or number of components will result in doubling that quantity.

In summary, additivity refers to the combination of quantities by addition, while extensivity relates to the scaling of a quantity with the size of a system. Additivity allows the simple addition of quantities, whereas extensivity determines how a quantity scales as the system size changes. While these concepts originated from physics, I use them for social systems here. However, social systems are non-additive and non-extensive. Here are some examples of non-extensive and non-additive systems in the context of social systems.

Imagine a group of high school students conducting a project in chemistry. Each group member contributes equally to completing the project. The overall output or outcome of the group project is the sum of the contributions of each member. In this case, the group dynamics is both additive and extensive because the contribution of each member to the project is equal and because the group's output scales with the number of members, and the contributions of each member can be summed. This situation is totally imaginary, as real social groups, from high school students to political parties, are non-additive and non-extensive. The contribution of the different members of the group will always be different. Probably only a few students will contribute significantly to the project's success. In addition, the situation is clearly non-extensive. Doubling the number of students may not double the quality of the project. Knowing something about group dynamics, we may even hypothesize that it will reduce the quality of the results.

Let us think about a different example. A team is working on a complex problem that requires collaboration and expertise from different domains. Each team member brings specialized knowledge and skills to the table, and their *interactions* and contributions are interdependent. In this case, the group dynamics is non-additive because the team's collective problem-solving ability is not merely the sum of the individual abilities. Furthermore, the group dynamic is non-extensive because doubling the number of team members may not necessarily double the problem-solving capability, due to coordination challenges and potential diminishing returns.

Tsallis entropy, named after the physicist Constantino Tsallis, is a generalization of the Shannon entropy and is used in certain contexts where traditional entropy measures may not be appropriate. Tsallis entropy is not additive in the same way as the traditional Shannon entropy. When considering a composite system consisting of independent subsystems, the total Tsallis entropy is not equal to the sum of the entropies of the subsystems. Indeed, the total Tsallis entropy of a composite system depends on the interaction between its subsystems. Regarding the idea of extensivity, the Tsallis entropy of a system is non-extensive when its entropy index q is different from 1. Extensivity in the context of entropy means that the entropy scales linearly with the size or extent of a system. However, Tsallis entropy generally violates this property and exhibits non-extensive behavior. Indeed, when $q \neq 1$, the Tsallis entropy does not scale linearly with the size of the system, and doubling the size of the system does not result in a doubling of the entropy.

Tsallis entropy is typically used when the underlying system exhibits non-extensive and non-additive behavior. In this context, an important component of this entropy measure is the entropy index "q." When $q > 1$, it emphasizes more frequent events in the distribution and is described as sub-additive and sub-extensive. When $q < 1$, it amplifies rare events (Tsallis et al., 1998a, 1998b).

Measuring the Entropy of Political Parties

Let me repeat the explanation given at the end of the previous section. The Tsallis entropy measure includes an entropy index q which is "responsible" for measuring the "non-additivity" of the system. If the entropic index is smaller than 1, it amplifies less probable cases, in our case, smaller political parties comprising the assembly/government. In contrast, if $q > 1$, more emphasis is given to the more common cases, in our case, bigger parties. In other words, if $q < 1$, low probabilities are enhanced, and if $q > 1$, high probabilities are enhanced (Tsallis et al., 1998a, 1998b, 2003). Put another way, $q < 1$ and $q > 1$ privilege the *rare* and the *frequent* events, respectively.

Now suppose that political wholes are different from the sum of their parts. Then we can better understand them using an entropy index that allows for non-extensivity and non-additivity. For the nine government compositions, I calculated the entropy of the National Assembly using three parameters: $q < 1$(i.e., $q = 0.2$), $q = 1$ (which coincides with the Shannon entropy), and $q > 1$ (i.e., $q = 2$). I hypothesized that if we are surprised by the difference between the expected and observed nationalism scores, then we should observe a statistically significant correlation between the surprise score (i.e., the difference between the expected and the observed score) and the entropy score of the composition with $q = 0.2$.

To test this hypothesis, I measured the correlation between the entropy scores and the gap between the observed score found by AI and the expected score calculated by averaging the parties' positions (expAVG). Using a common approach for calculating the prediction error, I computed the surprise involved in observing the actual score of the government using

$$\text{surprise} = (\text{observed} - \text{predicted})^2$$

Next, I computed the Pearson correlation between the surprise and the three entropy scores. The results are shown in Table 6.3.

We can see that the highest correlation was found for $q = 0.2$. The most important measure is the Bayes factor (BF), showing the odds in favor of our

Table 6.3 Correlation between the entropy measures and the surprise score

q index	r	BF	VS-MPR
0.2	0.81	16	16
1	0.77	10	10
2	0.76	9	9

hypothesis. As you can see, the Bayes factor for q = 0.2 is 16, while it is 10 and 9 for q = 1 and q = 2, respectively.

The results show that in a case where our expectations, which are based on naïve averaging, differ from the observed nationalism score, *the best model for explaining the size of the difference between our expectation and the observed nationalism score is the one using the entropy index emphasizing the importance of small parties.*

For example, the 1990–1994 assembly was *more nationalist than expected*. The explanation provided by the analysis is that the FKgP party, which holds a right-wing position, pushed the MDF, which is a more center-oriented party, to the right. The same holds for the first Orbán government, which was *less* nationalist than expected. The balancing force of the MSZP party, with its center-left orientation, seems to have had an important role in this surprising moderation.

The difference between the entropy scores with q = 0.2 and q = 1 is another measure we can use. It indicates the difference in uncertainty when amplifying the weight of the small political parties and the baseline, which is the uncertainty as measured by the Shannon entropy. This difference was strongly and positively correlated with the surprise score ($r = 0.78$, $p < 0.001$, VS-MPR = 10.69). The greater the influence of the small political parties, the greater our surprise over the government's position in which they were a part.

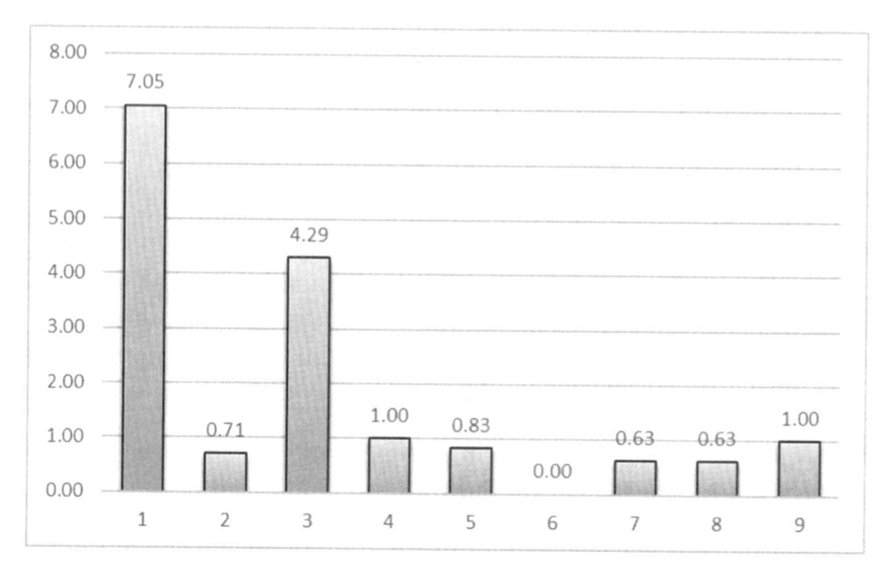

Fig. 6.1 Ratio between the two entropy scores for each government. *Source* Author

We can better understand this correlation by converting the entropy scores using the MAD approach and computing the ratio between the entropy scores with q = 0.2 and q = 2. The higher the score, the more important the small political parties are in explaining our surprise regarding the actual nationalism level. Figure 6.1 presents the ratio between the entropy scores for each government.

We see that the first government of Antall was characterized by the most significant weight of the smaller party. In contrast, the three last Orbán governments were an expected result given their simple composition from two right and far-right parties. What you see is what you get.

What if the Whole Is Different from the Sum of Its Parts?

In this chapter, I have discussed the nonlinear effect of the political parties comprising a government. In some cases, smaller parties may have a disproportionate influence on the whole, either by moderating the nationalism level of the government (e.g., the first Orbán government) or by moving the government toward a more nationalist position. While our focus may be biased toward the dominant parties and their leaders, the results presented in this paper draw our attention to the significant non-linear interactions associated with a government's political position. The disproportional influence of small political forces is not new and is nicely discussed by Taleb (2018). For example, when experiencing terror attacks by radical Islamists, some good souls remind us that most Muslims are not terrorists. It is a fact that the overwhelming majority of the Muslim population in the world has not been involved in any act of terrorism. However, what Taleb emphasizes is the disproportional influence of minorities. The majority may have no significance when small groups of zealous individuals take the lead.

While this understanding may be trivial, the development of models to explain and predict the political position of a political composition is less trivial, given the non-linear effect of their assembly. With its trivial limitations, this chapter does not intend to propose a complete model, only to demonstrate the importance and feasibility of using foundational scientific tools to get a better understanding of composite political structures.

The question is: So what? What have we learned about the dynamics of collectives and the ways that individuals may find their place within them? So far, I have discussed the crowd mostly as a blind herd of individuals showing homogeneity in their actions. Some crowds do indeed behave like a herd. In dictatorships, opposition forces are brutally silenced, resulting in a relatively homogenous crowd. In other contexts where variability exists, the system presents some degrees of freedom for the empowered individual. This chapter adds complexity to our understanding of crowds by showing that variability in the crowd's composition may have highly important consequences for the behavior of the whole. In totalitarian regimes, the crowd is led blindly by the dictator. However, in contexts where some variability is allowed, *small components of the whole may have a disproportionate influence on the behavior of the larger collective.*

This understanding gives us hope, as it emphasizes the importance of small organizations and the promise of non-linear interactions. For example, when playing a strategic game in political elections, the individual may play a more important role by *betting against his own crowd* and choosing a smaller party that may not fully represent his ideological beliefs, but may have more practical influence as a minority *within* the coalition. In the context of Orbán's government, voting for a small party that may moderate the nationalist inspiration of the leader may have more influence on the situation than voting for the biggest opposition party even though it has no chance of being part of the government.

We learn that, sometimes, the choice is not between good and bad but between bad and worse. In these cases, betting against the crowd may take a paradoxical turn. We may fail to bet against the crowd by supporting the oppositional crowd. But we may bet *against* the crowd by supporting a moderating force that can balance the system. Another important lesson is that variability is crucial for maintaining balance. Individuals who bet against the crowd may be empowered by understanding their contribution to the increased *variability* of the collective in which they are a part. Betting against the crowd does not necessarily involve a direct contrarian and antagonistic approach. Sometimes, soft power and a flexible mind may be the best non-stupid choice.

References

DW Documentary. (Undated). What causes monster waves? Accessed 7 Jan 2022. Available at https://www.youtube.com/watch?v=3wWXeSvvyUY&list=WL&index=7

Dysthe, K., Krogstad, H. E., & Müller, P. (2008). Oceanic rogue waves. *Annual Review of Fluid Mechanics, 40*, 287–310.

Gessen, M. (2021). The historians under attack for exploring Poland's role in the holocaust. Accessed 5 June 2023. Available at https://www.newyorker.com/news/our-columnists/the-historians-under-attack-for-exploring-polands-role-in-the-holocaust

Kingsley, P. (2019). A friend to Israel, and to Bigots: Viktor Orban's 'double game' on anti-Semitism. Accessed 5 June 2023. Available at https://www.nytimes.com/2019/05/14/world/europe/orban-hungary-antisemitism.html

Müller, M. (2021). For Polish nationalists, public enemy number one is not LGBT or Muslims, but the Westerner. Available at: https://notesfrompoland.com/2021/12/20/for-polish-nationalists-public-enemy-number-one-is-not-lgbt-or-muslims-but-the-westerner. Accessed 1 July 2023.

Neuman, Y. (2021). *How small social systems work: From soccer teams to jazz trios and families*. Springer Nature.

Neuman, Y., & Cohen, Y. (2023). AI for identifying social norm violation. *Scientific Reports, 13*(1), 8103.

Taleb, N. N. (2018). *Skin in the game: Hidden asymmetries in daily life*. Random House.

Thorpe, N. (2020). Hungary broke EU law by forcing out university, says European Court. Accessed 6 June 2023. Available at https://www.bbc.com/news/world-europe-54433398

Tsallis, C. (2014). An introduction to nonadditive entropies and a thermostatistical approach to inanimate and living matter. *Contemporary Physics, 55*(3), 179–197.

Tsallis, C., Mendes, R., & Plastino, A. R. (1998a). The role of constraints within generalized nonextensive statistics. *Physica A: Statistical Mechanics and Its Applications, 261*(3–4), 534–554.

Tsallis, C., et al. (1998b). The role of constraints within generalized nonextensive statistics. *Physica A: Statistical Mechanics and Its Applications, 261*, 534–554.

Tsallis, C., et al. (2003). Introduction to nonextensive statistical mechanics and thermodynamics. arXiv preprint cond-mat/0309093.

West, B. J. (2016). *Simplifying complexity: Life is uncertain, unfair and unequal*. Bentham Science Publishers.

Wikipedia A. (Undated). Government of Hungary. Accessed 20 July 2023. Available at https://en.m.wikipedia.org/wiki/Government_of_Hungary

Wikipedia B. (Undated). Elections in Hungary. Accessed 20 July 2023. Available at https://en.m.wikipedia.org/wiki/Elections_in_Hungary

7

Fixed Beliefs and Football Fandom: Unraveling the Dynamics of Collective Optimism

Introduction

> It is easy to imbue the mind of crowds with passing opinion but very difficult to implant therein a lasting belief. However, a belief of this latter description once established, it is equally difficult to uproot it. (Le Bon, 1895, p. 143)

Peter Pan (Barrie, 2014) is one of the classic books of all time. The book negatively describes one of its "heroes": Like all slaves to fixed ideas, it was a stupid beast. This refers to the crocodile remembered for biting Captain Hook's hand. However, this famous crocodile is not the only slave to a fixed idea. Luckily, I have not met any crocodiles, so I cannot generalize beyond this one to others of its kind. However, I am a human being who has met other human beings, both individuals and crowds, and I can testify personally that we are all prone to fixation—specifically, fixation on certain ideas. And when we are set on some particular idea, we will often be disillusioned too late.

I once met a former citizen of the USSR who told us how Stalin was worshipped. Not just respected or admired, but worshipped, as though he was God. Regardless of his atrocities, many people did indeed behave like slaves to a fixed idea that was disconnected from any reflective and critical thought. These people were the very expression of idol worshippers. From ancient times, Jewish monotheism mocked the idol worshippers who prayed to and made sacrifices to incompetent "gods." Paradoxically, the modern heirs of monotheism continuously violate the ancient imperative by performing

Y. Neuman, *Betting Against the Crowd*, https://doi.org/10.1007/978-3-031-52019-8_7

daily practices of idol worshipping. Fixation seems to be the rule rather than the exception.

It is remarkable that our minds, which strive to maintain a delicate balance between flexibility and stability, favor rigid stability when they become particles among other particles in a crowd. As I explained in the summary above, individuals and crowds seem to suffer from the same fixation, but being a particle rather than a thinking individual may amplify this tendency. Le Bon's observation, as quoted above, suggests that crowds may settle themselves on a certain belief from which it is difficult to move. In some cases, this is a depressing thought. Imagine a charismatic leader promising his followers that they will go to heaven if they accept his final solution to the problem. Hitler was such a leader, and his followers were a crowd that resisted any change of opinion till their final annihilation by the Allies and the Red Army. Some false (and "true") beliefs may resist change regardless of any conflicting evidence.

In the first chapter, I explained why facts should be emphasized over truth, specifically when dealing with the unpredictability of the crowd. A crowd can follow a "truth" that is eventually exposed as no more than a deadly illusion. In contrast, by definition, facts do not lend themselves easily to the devices of the madding crowd. Gaining a possible edge over the crowd requires close attention to facts rather than passionately seeking the truth, with its risk of leading to fixation and blindness.

How stable are a crowd's beliefs? One way to answer this question is by analyzing the stability of public opinions. As illustrated in the introduction, analyzing public opinions is a common practice, but it has its limitations. An opinion is just an opinion, and it is unclear whether it represents an obligation to engage in any specific action. Opinions are tricky representations of a crowd's beliefs, and *betting odds* are much better. After all, reporting about your beliefs for a national survey entails no obligation or cost, rather like the declaration of a Sicilian mafia boss that he firmly and truly believes in Santa Maria.

Declarations are always less convincing than compulsory behavior that involves putting your "skin in the game" (Taleb, 2018). For example, in football, your betting odds mean that your optimism concerning your favorite team is materially expressed in the money you are ready to pay and risk for your belief. Without some potential cost, there is no genuine expression of a belief or ideology. Therefore, in this chapter, I will test Le Bon's hypothesis by examining the dynamics of optimism in football, as expressed in how crowds bet on their favorite team.

First, I would like to show that the crowd's optimism and pessimism are difficult to change, as hypothesized by Le Bon. However, as the book focuses on the individual betting against the crowd, I will be making two more points. The first is that by understanding the crowd's delayed understanding of a situation and fixation on ideas, the individual may quickly bet against the crowd by updating her beliefs. This is an important point closely connected with the idea of scaling. Big things move more slowly, and once moved, they are less flexible when it comes to changing their course. The individual may bet against the crowd by using these differences in scale and the fact that she can act faster and more flexibly to a changing reality.

The second point is that we usually think about the behavior of individuals and collectives in terms of *skill* versus *luck*. I want to add a third ingredient to the complex soup of social dynamics by adding *trickery* into the equation. Hermes, the Greek goddess of trickery, will represent this component, alongside Lady Fortuna, representing luck, and Lady Athena, representing wisdom. By understanding Hermes' invisible hand, the individual may gain an advantage over the naïve crowd, which confuses skill, luck, and trickery.

Optimism and Pessimism in Football

My example concerns a fascinating case I have studied intensively in a previous book (Neuman, 2021). The 2015–2016 season of the English Premier League made a sensation. Leicester City Football Club, a team from a small town that no one believed in, except for some devoted fans, surprised everyone and climbed to the top of England's elite football league for the first time in its history. To understand just how sensational this was, we should recall that before the season began, the long-term odds of Leicester winning the league were 2000/1 (0.02%), while the odds for the favorite team, Manchester City, were 3/1 (25%). Against all odds, Leicester won, while other teams disappointed their fans. For example, Chelsea was one of the favorites to win the league but failed and ended at the bottom.

These gaps between expectations and realizations provide a wonderful case study for the persistence of beliefs as expressed in betting odds. Leicester entered the league with no positive expectations. It was destined to fail. In contrast, the betting odds for Chelsea indicated that the market was highly optimistic. Tracing the betting odds for Leicester and Chelsea throughout the season, we can determine whether beliefs changed or stayed relatively fixed. Figure 7.1 presents the betting odds in favor of Chelsea, a clear favorite for winning the league and disappointing this expectation. The betting odds have been transformed to increase visibility.

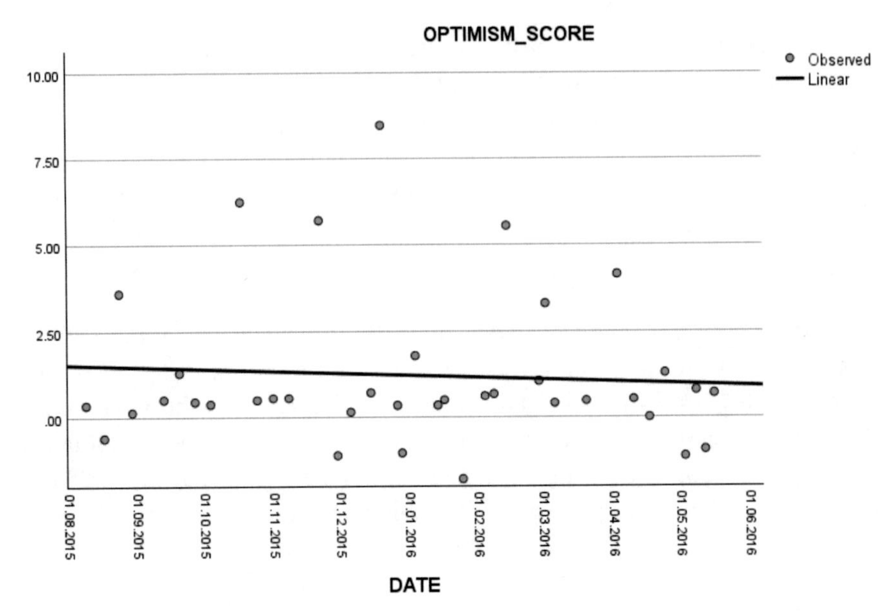

Fig. 7.1 Change in betting odds (i.e., optimism) in favor of Chelsea. *Source* Author

We know that this team disappointed its fans. Therefore, I expected to see a decreasing trend in the betting odds due to the increasing disappointment. A linear fit to the data shows no trend. Testing other trends failed, too. There seems to be no clearly decreasing trend that expresses the crowd's diminishing optimism. This is an interesting finding. The higher expectations from Chelsea were violated, so we might have expected the betting odds to reflect the crowd's disappointment. A clear trend of disappointment, as expressed in betting odds, could have been linear or exponential, but some decay should surely have been observed. However, such a clear trend is not observed.

What about the differences between the betting odds on consecutive occasions? These are shown in Fig. 7.2. Again, no clear trend is observed. The team's actual performance was not reflected in the betting odds. This is a single case but a clear one. Here, we have a team that disappointed the betting crowd, but the crowd's optimism, as reflected in the betting odds, did not show any decay. Optimism remained despite the evidence to the contrary. It seems that the crowd's optimism was quite resistant to the facts and behaved similarly to a religious "truth."

What about Leicester? The Cinderella story of the league? For a positive surprise, we might have expected an increasing trend of betting odds, indicating the crowd's understanding that magic was happening before its eyes. The results are shown in Fig. 7.3. In this case, we have a slightly increasing trend with a huge variability surrounding it.

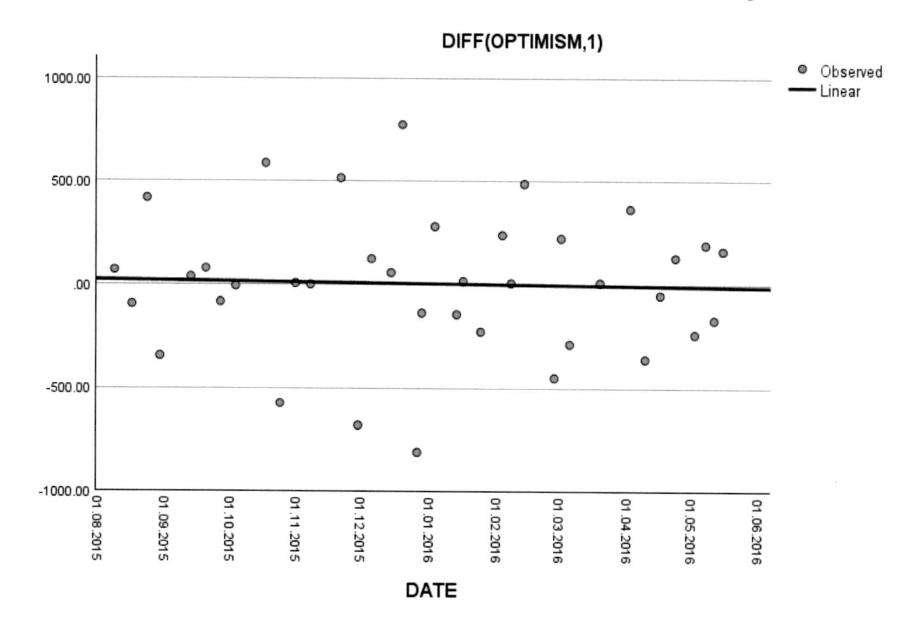

Fig. 7.2 Changes in betting odds as a function of time. *Source* Author

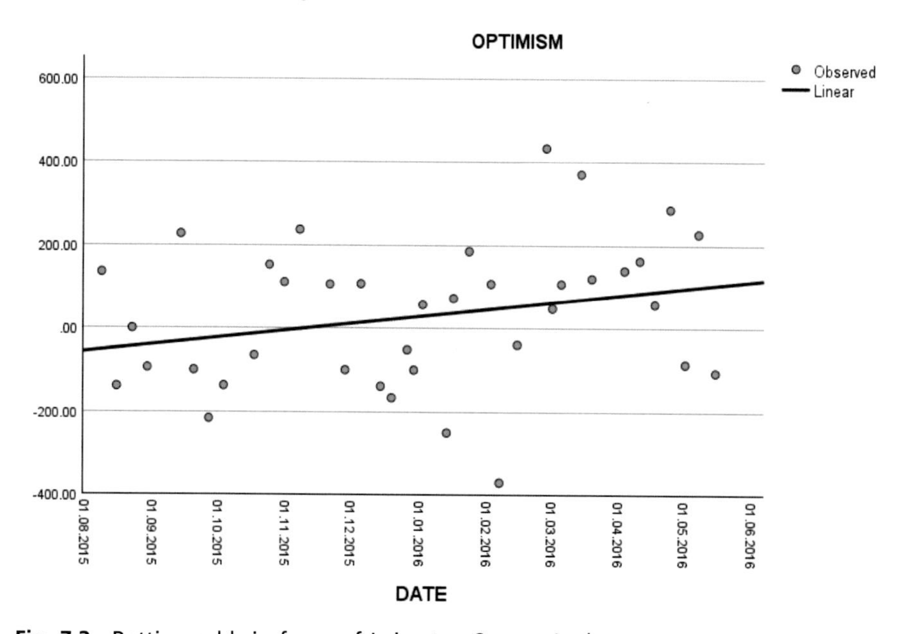

Fig. 7.3 Betting odds in favor of Leicester. *Source* Author

For the changes in the odds, the graph is shown in Fig. 7.4. It shows again that no clear trend is observed. This is a surprising result, too. Here, we have a Cinderella team coming from nowhere and paving its way to the top. However, the fixed beliefs associated with the newcomer seem to hold regardless of the undeniable fact: this team is a miracle!

So far, our analysis has been limited to two graphs presenting the betting odds for a positively surprising team and a negatively surprising (i.e., disappointing) team. It is remarkable how these extreme case studies of positive surprise and disappointment reflect how fixed a crowd's beliefs can be. Let us delve deeper into the data.

Given our enthusiasm for Cinderellas, an interesting question is whether the crowd expressing its optimism through the bookies' odds is sensitive to the positive surprise (i.e., Cinderella behavior) manifested by the team throughout the season, in the sense that it changes its degree of optimism/pessimism as a function of the team's positive violation of expectations (i.e., Cinderella) or negative violation of expectations (i.e., disappointment).Le Bon (1895) realized long ago that crowds have fixed ideas. Despite their capricious manner, I hypothesized that the crowd would be less impressed by Cinderella's behavior than by the confirmation of its expectations.

The crowd expresses its optimism by giving higher winning odds for one of the teams in a match. In the case of Leicester, and over the whole season,

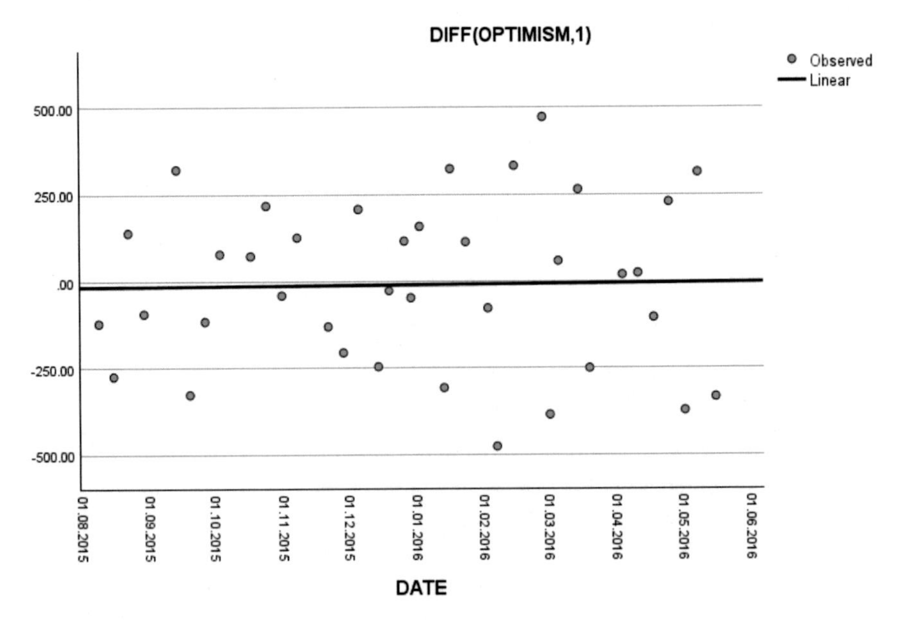

Fig. 7.4 Changes in betting odds for Leicester as a function of time. *Source* Author

the odds in favor of Leicester winning a game were

$$\text{Odds}_{\text{Leicester}} = 0.57/0.43 = 1.32$$

where the probability of winning the game was 0.57. Now, I analyzed all of Leicester's matches and asked what the probability was that the odds would favor Leicester given that, in the previous match, the team:

1. Confirmed the crowd's positive expectation (Positive Expectation Confirmation)
2. Negatively violated the crowd's positive expectations (Disappointment)
3. Confirmed the crowd's negative expectation (Negative Expectation Confirmation)
4. Positively violated the crowd's negative expectation (Cinderella).

The probability of betting in favor of Leicester, given a Positive Expectation Confirmation in the previous match, was $p = 0.61$; given Disappointment, it was $p = 0.75$; given a Negative Expectation Confirmation, it was $p = 0.33$; and given Cinderella, it was $p = 0.55$. A Cinderella game is one in which the team surprises us in a good way, despite the betting odds against it. This is a classic case of a positive surprise. Our expectations were negative, and the team won. It might have been expected that the betting odds in favor of Leicester would be higher than the baseline of winning set by the team's probability of winning in general ($p = 0.57$), given a nice surprise in the previous game. However, the probabilities were almost identical ($p = 0.55$ vs. $p = 0.57$).

Interestingly, the probability of betting for Leicester, given *disappointment*, was higher than the probability of betting for Leicester, given positive confirmation. Surprisingly, Cinderella's surprise did not change the crowd's beliefs. The following complementary Bayesian analysis aims to further enrich our understanding.

Using Jaynes's approach (Jaynes, 2003), we can represent the prior odds in favor of Leicester as

$$e(H) = 10 * \text{Log10}\left(\frac{P(H)}{P(-H)}\right) = 1.2$$

This means that taking the whole season's data, the odds in favor of Leicester were positive but only slightly higher than the odds expected if we assume an equal probability of winning or losing (i.e., odds of "1").

Now, we can ask whether the evidence that Leicester surprised and, against all odds, won the game improved the betting odds (the crowd's optimism) in the following match. When taking the evidence that the team behaved like Cinderella in the previous match, we see that, based on the data, the crowd's expectations have been updated as follows:

$$e(H) + 10 * \log 10\left(\frac{P(E/H)}{P(E/-H)}\right) = 1.14$$

In this case, the result is 1.14, meaning that, given the evidence that the team expressed *positive surprise* in the previous match, it entails nothing positive to the gamblers setting the odds. In fact, the betting odds were even lower! So, observing Leicester's surprising success had little or no influence on the crowd. It is as if a failed high school student were still considered a failure even after proving to his teachers that they were wrong. This is the meaning of a stigma.

Surprisingly, it was a disappointment from the team that was most informative, moving the crowd's prior beliefs from 1.2 to the updated belief of 1.57. Disappointment seemed to have a reverse effect: instead of lowering the betting odds in favor of Leicester, it increased them. Positive Expectations Confirmation slightly strengthened the belief in the team and moved the odds of winning from 1.2 to 1.27. Nothing impressive. The crowd's expectations regarding a team seem to be influenced by *prior beliefs* rather than by surprising behavior and violation of expectations. In the case of increasing optimism that follows disappointment, it may be that the enthusiastic Leicester fans are those betting for the team as a kind of supportive statement rather than as a result of fact-checking.

The analysis of a clear and extreme case shows how fixed the crowd remains, even if it has to pay for being fixed. Those who present ungrounded optimism about the potential of crowds to change seem to be voicing an ideologically biased opinion and nothing more. Although my analysis is extremely limited, it has its lesson. If you believe a given crowd can change, you need to justify your belief, as it seems that crowds are settled on their prior "opinions." Football fans may not be representative of all crowds, and in some cases, they clearly respond to the changing status of teams. However, in extreme cases of expectation violation, the crowd may behave like a slave to a fixed idea, as would almost naturally be expected from crocodiles and human beings.

So far, our discussion has focussed on how a crowd's beliefs can become fixed. However, Cinderella or the underdog phenomenon is a great opportunity to deepen our understanding of chance, which has bothered us since the beginning of the book. In particular, it bothered us in the context of our freedom as individuals to bet against the crowd. Here, we can deepen our understanding by adding another interesting factor to the equation.

Cinderella Teams and Hermes' Invisible Hand

According to some naïve conceptions, "winning against all odds" is winning against chance, as trivially implied by the phrase itself. But this phrase represents a naïve and over-simplistic approach. When an individual decides to bet against the crowd, she may think that her chances are so low that she is betting against *all* odds. However, the odds are based on our partial and limited knowledge, such as the football team's previous reputation. Odds, like entropy, are not an ontic but an epistemic concept, and the two are too easily confused. Therefore, and in contrast with naïve conceptions, the odds represent our *uncertainty* and not our *certainty*.

Winning against all odds means that, regardless of our idea of certainty through partial understanding, Lady Fortuna is kind enough to allow us some degrees of freedom and the joys associated with them. Therefore, being a Cinderella team may result from (1) "chance" or (2) the intrinsic skill of an underdog team winning against all odds. Or indeed, it can be a mixture of the two in complex situations. However, like other stories of success, Cinderella stories love to emphasize the hidden talent they reveal almost magically to the world. So, let me explain a little better what is meant by being a Cinderella.

Cinderella stories describe situations where "competitors achieve far greater success than would reasonably have been best expected." (Wikipedia). We love such stories as we can identify with the struggle (Keinan et al., 2010) of hard workers (Wang, 2016) winning against all odds by exposing their hidden talent. In a complex world that is unfair, uncertain, and unequal (West, 2016), believing in a hidden talent just waiting to be exposed to the world is a comforting idea.

Some games, such as dice, are pure games of chance, while others, like chess, are skill-based (Duersch et al., 2020). The scientific papers dealing with skill vs. luck make a simple assumption: a game is about interactions between Lady Fortuna and Lady Athena, the goddess of wisdom. However, real life is not *only* about skill, wisdom, and luck, because tricksters, whose mythological goddess and patron was Hermes, are also an inherent part of our natural world (Stevens, 2016).

In this context, we may ask two interesting questions. First, is Cinderella-like *behavior* indicative of talent? Here, I am not discussing a one-shot behavior where a team surprised us for good but a consistent pattern of surprise. This pattern may indicate a hidden talent we might have missed. It is also possible that violating our expectations has nothing to do with talent and that this surprise only indicates our failure to predict. Therefore, Cinderella-like behavior seems to result entirely from Lady Fortuna's invisible hand. *Via negativa*, a counterintuitive hypothesis, suggests that the relative lack of Cinderella-like behavior indicates tricksters and Hermes' invisible hand. Therefore, my next question is whether Cinderella-like behavior is associated with trickery, as in match-fixing. To answer these questions, I analyzed two famous and well-documented cases of match-fixing and show that the lack of Cinderella behavior probably indicates a lack of randomness and may serve as a good indication of Hermes's invisible hand. This deeper understanding of chance, skill, and trickery may help us to find new ways of betting against the crowd.

Lady Fortuna's Invisible Hand

I analyzed the data of the 2015–2016 premiere league using a straightforward method. First, I used the bookies' odds for each match. If the odds were against the team and it won or gained a draw, the match would have been marked as Cinderella. For each team, I measured the percentage of matches in which the team positively surprised us and named this score "Cinderella." Figure 7.5 presents the correlation between the Cinderella score of a team (X-axis) and the points gained by the team at the end of the season (Y-axis).

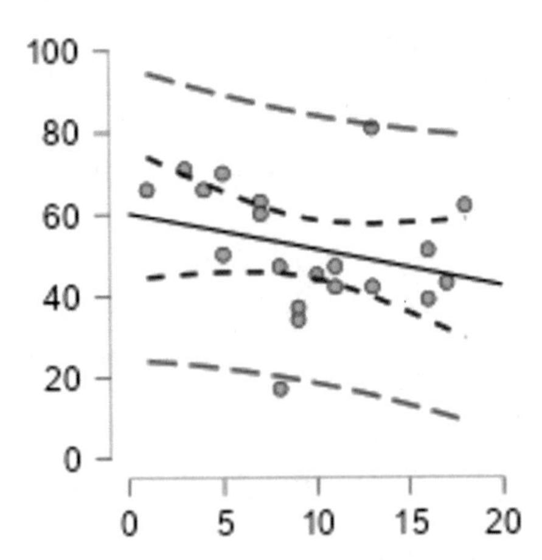

Fig. 7.5 Correlation between the Cinderella score and success at the end of the season

Using Spearman's rank correlation coefficient rho, the correlation is not statistically significant but mildly negative. Therefore, Cinderella-like behavior is, in the best cases, not associated with talent and, in the worst case, negatively associated with it. This is a painful finding. We all love Cinderella stories, and it may be extremely disappointing to learn that the Cinderella effect is no more than noise or uncertainty. What we gain from this is that, not only do we become more reflective and critical of the ability of crowds to change in time, but we also learn that the crowd's mind may be subject to manipulations other than its own inherently biased beliefs. Deciphering talent, skill, and wisdom from pure noise may give the individual an edge over the crowd seeking comfort in its illusions.

To test this idea further, I used the Cinderella scores of each team and computed the median of the absolute deviations from the median (MAD). I measured how far each Cinderella score was from the median in terms of MAD, and for each team, I identified the highest deviation. This score represents the *highest positive surprise expressed by the team.* I found that this score was *negatively correlated* with the overall points gained by the team at the end of the season (Spearman's rho $= -0.62$, $p < 0.001$). This means the team's best positive surprise *negatively* correlates with its success. Being surprising in a good way is not a good indication of a hidden talent. The long hand of Lady Fortuna seems to reach even our most cherished fantasies.

The hypothesis emerging from the last result is that expressing the "underdog virtue" to its highest level may be a sign of randomness rather than a sign of a real virtue. If this hypothesis is grounded, we should find fewer Cinderellas among leagues and teams known for corruption and match-fixing. After all, match-fixing, as the name suggests, reduces Lady Fortuna's influence and involves the invisible hand of Hermes.

Hermes' Invisible Hand

To test this hypothesis, I identified two classic cases of corruption and the teams involved. The first concerns the Turkish League (Yilmaz et al., 2019). During the 2010–2011 season, the Turkish football league was marred by a match-fixing scandal that had significant repercussions for Turkish football. The scandal involved match-fixing allegations, bribery, and manipulation of games in the top-tier Süper Lig. The investigation focused on several high-profile clubs and officials. *Fenerbahçe* and *Trabzonspor*, which were at the top of the league, were primarily associated with corruption; therefore, I analyzed the behavior of these two teams. The second case, the Calciopoli scandal (Migali et al., 2016), was a famous sports scandal in Italy in 2005–2006, where the elite team Juventus that won the season was punished for corruption. Therefore, I selected Juventus as the third team for analysis.

At this point, I had three teams known for corruption. Next, I compared the Cinderella scores of two clusters: the three genuine winners of the Premiere league I discussed before, and the Cinderella scores of the three corrupt teams. The Cinderella scores of the cheaters (i.e., exploiting the hand of Hermes) and the real winners are presented in the box plots in Fig. 7.6.

We can see that cheaters present Cinderella-like behavior to a lesser extent. Put differently; genuine winners present more Cinderella-like behavior. Recall that I hypothesized that Cinderella-like behavior indicates Lady Fortuna's invisible hand. The above graph supports my hypothesis. It means that fair games are characterized by more randomness than we may want to believe.

Next, I merged the percentage of Cinderella cases and cases where the odds favored the team and it lost or gained a draw (i.e., Disappointment). These are cases of *surprise*. Figure 7.7 compares the surprising behavior displayed by cheaters and real winners.

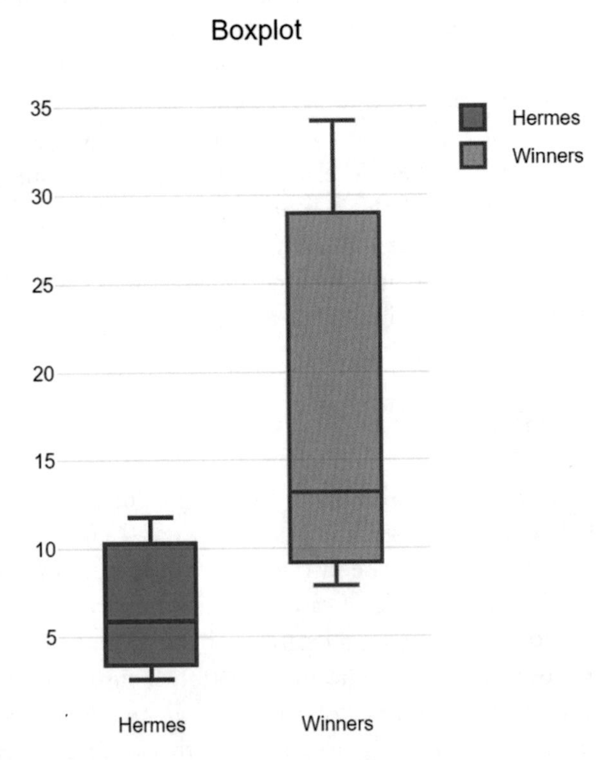

Fig. 7.6 Cinderella scores for cheaters versus winners

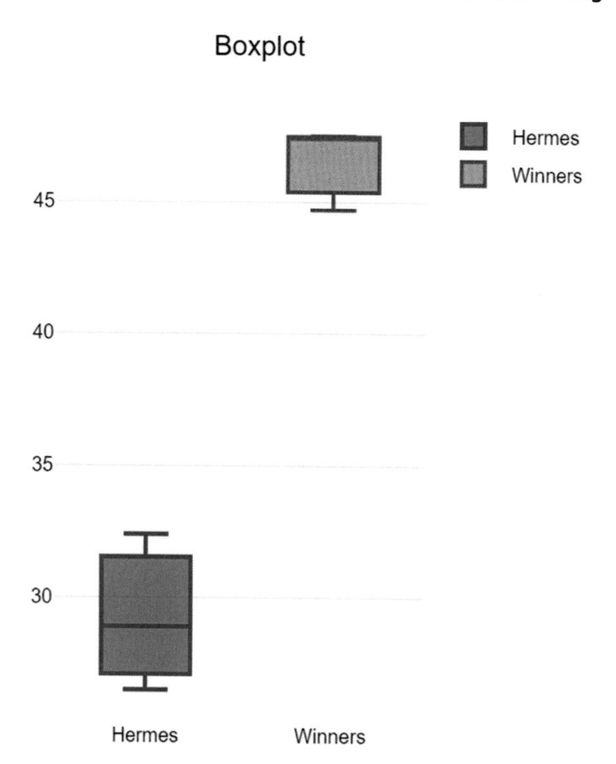

Fig. 7.7 Percentages of surprising matches for honest and match-fixing teams (i.e., winners)

We can see that surprise was lower among the cheaters, and for good reason. Match-fixing reduces the influence of Lady Fortuna and allows Hermes to take control. What are the implications of this understanding for the individual betting against the crowd? The most important implication is that chance is much more prevalent in real life than we may want to believe. In cases where the power of chance is diminished, we should always examine the hypothesis that it is Hermes' invisible hand that we are seeing.

"Knowledge Is Safety"

It has been argued that optimism and pessimism—expecting a positive or negative future—are two distinct modes of thinking grounded in different neural mechanisms (Hecht, 2013). It has also been argued that: "The principal differences between the two are: (a) Selective attention and information processing. (b) A belief (or lack thereof) that one has the power to influence relevant situations, events, and relationships (i.e., locus of control). (c) The general schema one holds for interpreting personal events (i.e., attribution style)". (ibid., p. 174).

Selective attention is evident when we focus our attention on things that we would like to see and ignore others that disturb us. In this context, fact-checking must involve checking all the facts regardless of whether we like them. The observation that Chelsea did not perform as expected is a fact. It is a measurable fact. However, the crowd's optimism did not significantly change. Like the audience in a magic show, we are all prone to take part in collective illusions that have a joyful aspect to them. Changing our beliefs is energetically demanding and may be emotionally painful. Crowds seem to avoid these difficulties. Betting against the crowd requires careful attention to how the crowd's attention is consistently biased. The individual betting against Leicester, by quickly adjusting his beliefs, could have profited much more than those who were late to realize it was a Cinderella team.

The second aspect of optimism and pessimism is our sense of control. Between an inflated sense of omnipotence or impotence, there is a wide range of context-dependent possibilities. The betting crowd may believe that it can influence the results of a match by supporting its team and going against the other. This is wishful thinking, but it works to a certain extent. A team playing a home match is more likely to win, but there is a limit to the extent that playing at home can benefit a team. Moreover, the crowd may attribute success to its team's talent and the fans' support and attribute failure to outside circumstances such as a biased judge or an away match. This is all psychology, which, interestingly, lacks the important aspect of uncertainty in the behavior of the teams themselves. Like other crowds, football fans may be biased by a lack of understanding and wishful thinking, an unrealistic sense of control, and ungrounded emotions.

Another paper dealing with optimism suggests that, on the individual level, optimism is associated with "the anterior cingulate cortex (ACC), involved in imagining the future and processing of self-referential information; and the inferior frontal gyrus (IFG), involved in response inhibition and processing relevant cues" (Erthal et al., 2021, p. 895).

Inhibiting responses and imagining the future may be important traits for suppressing impulsive behavior, but they may also suppress incoming evidence violating our prior expectations. This hypothesis is further supported by a study (Liu et al., 2023) arguing that "in adult brains, both domain-specific and domain-general regions encode violation-of-expectation." Therefore, violations of expectations are processed by complex neural processes that cannot be understood out of context. The crowd's level of behavior probably enhances this complexity. Violations of expectations are crucial for the critical individual betting against the crowd.

The above discussion brings us back to statistics and possible ways for dealing scientifically with our ignorance and uncertainty. In this context, and despite our fascination with Cinderellas, it seems that, when examined throughout the season, Cinderella behavior expresses our uncertainty and has nothing to do with hidden talent or an intrinsic magic sauce leading to success. A team, such as Leicester, presented fewer cases of Cinderella than West Ham (34.2% vs. 47.4% respectively) ranked seventh in the final table. In itself, beating the odds throughout the season is not a clear sign of any intrinsic virtue or Lady Athena's invisible hand. Cinderella teams showing their success at the end of the season are rare cases that reveal our ignorance, as evident from other teams that beat the odds throughout the season without reaching the top of the league. The most important insight one may gain from my analysis is that Cinderella-like behavior can be attributed for the most part to the invisible hand of Lady Fortuna, and its relative absence to the invisible hand of Hermes.

The implications for the individual seeking to bet against the crowd are clear. The crowd may tend to be fixed on an idea, even more than the individual, because self-reinforcement of beliefs, wishful thinking, and repression of critical thinking may be more enhanced in the context of the collective. Such fixed ideas may be used by the individual with the time advantage on his side. The individual may bet against the crowd by identifying persistent and ungrounded beliefs. Moreover, the crowd will be unaware of the three forces of Lady Fortuna, Lady Athena, and Hermes, the trickster. Observing the different forces and their relative influence may give the individual a clear edge. Finally, seeking evidence (i.e., facts) instead of the "truth" and flexibly updating beliefs may give the individual another edge. These directions may hopefully save us from behaving like Peter Pan's stupid crocodile.

References

Barrie, J. M. (2014). *Peter Pan.* MacMillan (Collector's Library).

Duersch, P., Lambrecht, M., & Oechssler, J. (2020). Measuring skill and chance in games. *European Economic Review, 127*, 103472.

Erthal, F., Bastos, A., Vilete, L., Oliveira, L., Pereira, M., Mendlowicz, M., Volchan, E., & Figueira, I. (2021). Unveiling the neural underpinnings of optimism: A systematic review. *Cognitive, Affective, & Behavioral Neuroscience, 21*, 895–916.

Hecht, D. (2013). The neural basis of optimism and pessimism. *Experimental Neurobiology, 22*(3), 173.

Houdini, H. (1906). *The right way to do wrong*. The Barla Press.
Jaynes, E. T. (2003). *Probability theory: The logic of science*. Cambridge University Press.
Keinan, A., Avery, J., & Paharia, N. (2010). Capitalizing on the underdog effect. *Harvard Business Review, 88*(11), 32–32.
Le Bon, G. (1895). *The crowd: A study of the popular mind*. T. Fisher Unwin.
Liu, S., Lydic, K., & Saxe, R. (2023). Using fMRI to study the neural basis of violation-of-expectation. *Journal of Vision, 23*(9), 4925–4925.
Migali, G., Buraimo, B., & Simmons, R. (2016). An analysis of consumer response to corruption: Italy's *Calciopoli* scandal. *Oxford Bulletin of Economics and Statistics, 78*(1), 22–41.
Neuman, Y. (2021). *How small social systems work: From soccer teams to jazz trios and families*. Springer Nature.
Stevens, M. (2016). *Cheats and deceits: How animals and plants exploit and mislead*. Oxford University Press.
Taleb, N. N. (2018). *Skin in the game: Hidden asymmetries in daily life*. Random House.
West, B. J. (2016). *Simplifying complexity: Life is uncertain, unfair, and unequal*. Bentham Science Publishers.
Wang, O. (2016). 3. Everybody loves an underdog: Learning from Linsanity. In *Asian American sporting cultures* (pp. 75–101). New York University Press.
Yilmaz, S., Manoli, A. E., & Antonopoulos, G. A. (2019). An anatomy of Turkish football match-fixing. *Trends in Organized Crime, 22*(4), 375–393.

8

Contrarian Strategies: Capitalizing on the Limits of Exponential Growth in Financial Markets

Introduction

In the previous chapters, I discussed crowd behavior. Here, I would like to delve deeper into herding behavior and examine its meaning in another financial context. This chapter therefore corresponds to previous chapters, starting from Chap. 1, in which I discussed exponential growth, through to those chapters in which I discussed constraints and prediction. So, let's begin by explaining herding behavior.

Herding is a form of social behavior that involves "the alignment of the thoughts or behaviors of individuals in a group (herd) through local interaction and without centralized coordination" (Raafat et al., 2009, p. 20). The key idea here is that there is no leader in the herd. The behavior emerges from local interactions that lead to a well-defined behavior. For example, the term "bull market" is used in financial markets to describe a prolonged period of rising prices. Why a bull? Because a bull tends to run forward with its horns raised, symbolizing an upward or rising market. This behavior is associated with strength, optimism, and a bullish outlook. When the bulls run forward, they do it with no leader. If you have ever watched the traditional event in Pamplona, Spain, where bulls run after people who enjoy the thrill, you would be hard put to recognize any leader among the bulls. Bulls and investors in a bull market seem to move in this way.

What explains this behavior? Several mechanisms have been proposed to explain this behavior, such as the existence of a positive feedback loop (Kameda & Hastie, 2015) that explains herding in contexts ranging from crime (ibid) to financial markets (Litimi, 2017; Zhang et al., 2021). Another

Y. Neuman, *Betting Against the Crowd*, https://doi.org/10.1007/978-3-031-52019-8_8

mechanism explaining herding is the spread of information in a network (Yook & Kim, 2008). As Pastor-Satorras et al. (2015) show, the spread of information follows exponential growth.

The two explanations of a feedback loop and the spread of information go hand in hand. For example, a shoeshine boy shines the shoes of a rich trader who happens to be speaking with a colleague about the roaring market and the enormous opportunities for young people to get rich. The shoeshine boy is a hub that spreads gossip. On the same day, he shares this gossip with fifty clients. Each of these clients shares the gossip with his friends, and the gossip spreads exponentially, or like wildfire, to use common parlance. This form of information-spreading, or more accurately, gossip-spreading, leads to a positive feedback loop. In the stock market, a positive feedback loop describes a situation where rising prices lead to further buying, which drives prices even higher. This creates a self-reinforcing cycle where positive price movements encourage more buying, pushing prices up.

The idea of a feedback loop is one of the most basic and powerful ideas in every thinker's arsenal. A feedback loop can be positive and create a self-reinforcing cycle. However, a feedback loop can also moderate the process by regulating the system toward a point of equilibrium. Despite the simplicity of these ideas, I have found them to have an enormous explanatory and predictive force. For example, in one of my papers (Neuman & Cohen, 2022), I used this idea to build an intelligent system that could identify a change in emotion in conversation, specifically in Mandarin Chinese. But let me return to herding.

There is an interesting aspect of herding. Herd behavior, evident in various fields, from ant escape behavior to financial markets, involves *symmetry breaking*. Symmetry is a foundational idea in the sciences and the arts. Symmetry breaking can be easily illustrated in the context of financial markets. Let us assume that we analyze a time series of stock prices. We segment the time series into blocks of three consecutive prices, convert them into ordinal patterns, as explained before, and identify monotonic increasing and decreasing patterns. For example, the sequence of prices {170, 940, 12,000} is represented as the monotonic increasing pattern {0, 1, 2} and the sequence {240, 9, 3} as the monotonic decreasing pattern {2, 1, 0}. As shown in Fig. 8.1, these patterns are mirror images of each other.

The probability of observing the monotonic increasing and decreasing patterns is equal in a fully symmetric and random market. In a random market, and in the long run, entropy is maximized, and monotonic increasing and decreasing patterns appear with the same probability. However, suppose

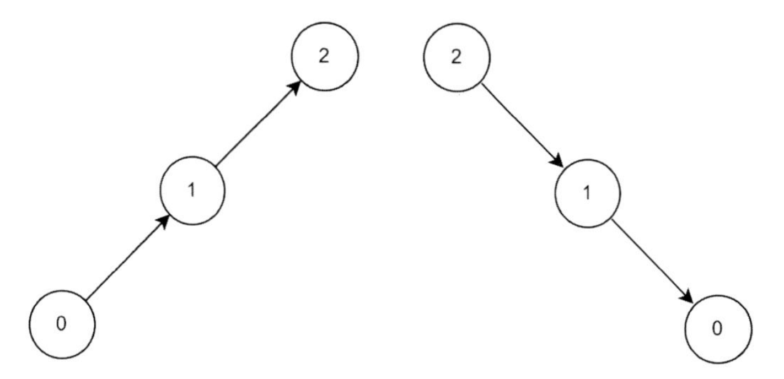

Fig. 8.1 The symmetry of monotonic increasing and decreasing ordinal patterns

we observe a bull market that moves the prices upward and forms an exponential growth trend. This dynamic can be expressed in terms of symmetry breaking because there is a higher probability of the monotonic increasing pattern.

In this context, measures of symmetry and asymmetry naturally suggest themselves for modeling herd behavior. But why exactly should we care about symmetry and symmetry breaking? Think about a trader who joins the market when he observes exponential growth. He joins the herd to ride the wave of success and gain a nice profit. However, he remembers the lesson from some of the previous chapters. Exponential growth is finite, and prediction is risky. Therefore, the trader wants to predict *trend reversal* so that he can leave the market in good time and avoid losses. In other words, the trader jumps on the wave of success but wants to quit when a changing trend approaches. This is a nice example of joining the herd, but betting against it and leaving the sinking ship well in advance.

In a paper I published in the *Journal of Statistical Mechanics* (Neuman & Cohen, 2023), I hypothesized that a transition from an exponential trend could be predicted by (1) the amount of asymmetry in the system and (2) the expected number of steps to symmetry, representing the time to relaxation and the return to equilibrium.

Let me explain these two points as they illustrate a repeating theme in this book: *profound simplicity*. The first phase of the process is to identify an exponential growth of prices in a given time window. Our trader can jump on the wave if the trend shows exponential growth. Otherwise, he avoids joining the market. This is the context of betting with the crowd. However, once on the wave, he must be on the lookout for a trend reversal that might throw him off the wave. In this context, betting against the crowd means recognizing a tipping point before the herd. The approach I will present aims to provide him with an indication that a trend reversal is approaching. First, I present the exact procedure that I used for identifying exponential growth. This is quite a technical section; the reader may skip it without significantly losing information.

Identifying a Significant Exponential Growth

First, I segmented the time series of stock prices using a sliding window of 12 observations and a time delay of 1. This means that we observe the average monthly price of the stock over a 12 month period. Then, we shift the window one step to the right. This procedure results in a sequence of 12 observations. Our attention is always focused on a time frame of 12 observations. This is the time frame we use to decide whether to join the market and also to predict an approaching trend reversal.

Second, using a logarithmic scale, I transformed the observations (i.e., prices) in each window. Many natural and economic phenomena exhibit exponential growth or decay patterns, from the rate and cost of DNA sequencing (McNamee & Ledley, 2012) to the financialization of art, where prices have grown exponentially through herd behavior (Emmert, 2018), and up to the Bitcoin market, where "The positive feedback effect between investors promotes the exponential growth in prices, and the herding effect eventually leads to a bubble" (Xiong et al., 2020, p. 4). Taking the logarithm of the dependent variable, these exponential relationships can be transformed into linear ones. Linear relationships are generally easier to analyze and interpret, allowing for the application of various statistical techniques that assume linearity. Therefore, I used the log-transformed prices as my dependent variable and the timeline of 12 observations (1–12) as the independent variable.

Next, I fitted a linear regression function to each window of 12 observations. The independent variable is the running index of the twelve discrete time points (1–12), and the dependent variable is the log-transformed score of the stock price. If the linear regression fits the data, *it implies that the trend is exponential*.

How do you decide whether a linear model fits the data? To evaluate the fit of the linear regression, I used Cohen's f^2 measure of *effect size* (Cohen, 1988), which is defined as:

$$f^2 = \frac{R^2}{1 - R^2}$$

where R^2 is a measure of the explained variance. I then used Cohen's common, albeit arbitrary, standards for effect size as *small*, *medium*, or *big* to define a new binary measure of exponential growth: SIGSLOP (i.e., significant slope). In other words, I used the explained variance of the model and used a convention to decide whether the "effect size" was small, medium, or big. The decision rule is simple: if f^2 is medium or higher (> 0.15) and the regression coefficient is positive, then the growth is considered exponential, and the slope is marked as significant. In this case, SIGSLOP scores "1" otherwise SIGSLOP = −1. Our trader joins the market only if SIGSLOP = 1.

Let me sum up. I observe a time frame of 12 observations and decide whether the stock price increase is positive and exponential. If it is positive and exponential (i.e., SIGSLOP = 1), the trader joins the market and starts trading.

Leaving the Ship in Good Time

The challenge facing the trader who joined the market when SIGSLOP = 1 is to identify in advance any transition from exponential growth. It is the same challenge facing rats that may need to abandon a sinking ship. While the rats have a real-time indication of the problem, our trader faces a more difficult challenge. She seeks to predict an approaching transition from exponential growth so that she can leave the market in time.

For all analyses from now on, we shall attempt to predict a transition from significant exponential growth to trend reversal (SIGSLOP = − 1) in a forthcoming time window up to six shifts from our current location (τ = 1–6). This means that, when a trader observes an exponential price growth in 12 consecutive data points (i.e., months), she joins the market. However, she uses a moving window of 12 data points and shifts it from one to six places to the right to understand whether a trend reversal is expected in the future. When the window is shifted one place to the right, there is an overlap of 11 months, and the prediction is for one month ahead only. When the window is shifted two months to the right, the prediction is for a time window of 12 months that overlaps with 10 months of the previous year and provides us with a prediction for two months ahead, and so on.

My next task was to provide the trader with a tool that can warn her in advance about an approaching trend reversal. To measure the asymmetry in a time series, I represented the time series of prices as a sequence of ordinal patterns, as explained in the previous chapters. Next, I proposed two measures that can be used to predict a transition from exponential growth. These are signs that can be used for prediction.

Signs of Reversal

The first sign of an approaching trend reversal is the system's asymmetry. This is computed as follows. For each sliding window of 12 observations (i.e., the stock's average monthly price for each of the 12 months), I measured the monotonic increasing and monotonic decreasing permutations: π_1 and π_6 (i.e., {0, 1, 2} vs. {2, 1, 0}, respectively). This is a simple procedure. I measured the percentage of monotonic increasing and decreasing patterns in the time window. The difference in the probabilities of these permutations has been used by Bandt (2020) to define a metric of *irreversibility*. Drawing on this work, I measured the *ratio* between the probability of the monotonic increasing permutation π_1 and the probability of its complementary mirror image (i.e., π_6):

$$\text{ASY} = \frac{P(\pi 1)}{P(\pi 6)}$$

This measure, called ASY for asymmetry, represents how much the increasing monotonic permutation π_1 dominates a given time series segment compared to π_6, its "mirror" permutation. The higher the ratio, the more asymmetry is observed in the system, meaning *the more dominant the upward trend*. Metaphorically speaking, the measure of asymmetry represents the force pushing the market upward. The stronger the force, the more likely the persistence of the upward trend and the longer it may last. An asymmetry measure that gets closer and closer to 1 means that the system is maximizing its entropy (i.e., symmetry) and approaching a point where the upward trend balances with the downward trend.

The second measure I used is StS (i.e., steps to symmetry). I assumed that irreversible exponential growth is time-limited by available energy resources, as is evident in many systems (e.g., West, 2018). I sought to predict when the system would return to equilibrium, full symmetry, and maximal entropy. Therefore, I used the exponential decay function

$$\text{P(t)} = \text{P}_0\text{e}^{-\text{rt}}$$

where P_0 is the initial value, r is the decay rate, and P(t) is the final amount. Specifically, I wanted to identify the *relaxation time* to the state where the system's final amount of asymmetry is in "equilibrium." In other words, I wanted to predict the *discrete time steps the system would take to reach full symmetry*, when the ratio between the probabilities of the monotonic increasing and the monotonic decreasing patterns is 1. Therefore, I used the following equation:

$$t = -\log(P(t)/P_0)/r$$

In this case, P_0 is the ASY value of the time series segment being analyzed. P(t) is the final amount of asymmetry in the system where $p(\pi_1) = p(\pi_6)$, and r is the estimated decay rate. The StS measure is therefore defined as

$$StS = -\log(1/ASY)/r$$

Let me summarize and explain.

The justification for the two measures is as follows. The first measure of asymmetry is the ratio between the probabilities of the monotonic increasing and the decreasing permutations in a given time series window. The higher the metric, the higher the asymmetry. The higher the asymmetry, the stronger the force pushing the system upward. The stronger the force pushing the market upward, the longer the expected trend. The second metric, StS, measures the relaxation time of the system given the initial amount of asymmetry. It approximates the time remaining before reaching full symmetry.

I hypothesized that the herd behavior pushing the market upward and the StS measure, indicating when the stock market is expected to relax and return to equilibrium, may be used to predict a transition from exponential growth. I further hypothesized that using the two metrics as predictors in a machine-learning model would significantly improve our prediction of a trend reversal. The reason for using machine-learning models is trivial. These are the best classification and prediction models that we currently have.

The Experiment

I used ASY and StS as variables in a machine-learning model to predict a transition from exponential growth to trend reversal (i.e., SIGSLOP = − 1). So, I provided the model with two scores and asked it to predict whether the next time window would involve a trend reversal from exponential growth. But how do we know whether these variables improve our prediction of a trend reversal? And in comparison with what? To find out whether our signals have any good informative value, we must compare our prediction to some yardstick or baseline. One possible baseline for prediction is the conditional probability of observing a trend reversal in a time window, given that we observed exponential growth in the previous time window (SIGSLOP = 1). In other words, I compared the performance of the predictive models with predictions given by the probability of observing SIGSLOP = −1 in window $W_{i+\tau}$, given that the current window (i.e., W_i) was characterized by exponential growth (SIGSLOP = 1). This conditional probability was the baseline used to evaluate the performance of the machine-learning classifier.

We analyzed three datasets using the same methodology and procedure. However, we present here the results for the Dow Jones only, and the interested reader may consult our paper for the other datasets. We first analyzed the monthly measurements for the Dow Jones Industrial Stock Price Index (1915–1968). Figure 8.2 presents the time series of the prices.

The context of the analysis is straightforward. Observing N consequent measurements (e.g., average monthly values of the DJ index), I identified time segments of exponential growth (i.e., SIGSLOP = 1) and used ASY and StS to predict trend reversal in a forthcoming time window.

The Dow Jones dataset included 357 windows labeled "1" (56%). These are windows where we observe exponential growth in the stock's price. The other 268 windows were labeled "−1" (42%), and these time windows indicate the non-existence of exponential growth. The rest of the time windows were missing cases.

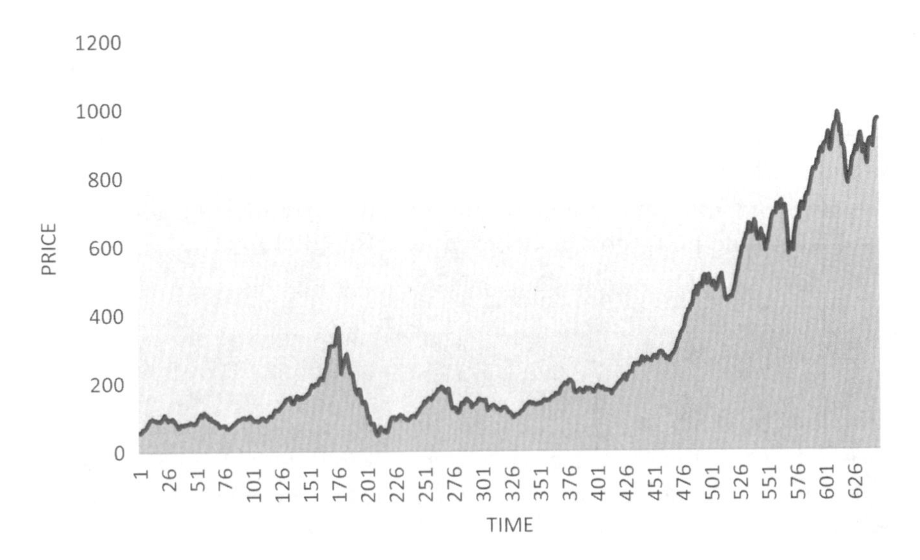

Fig. 8.2 The Dow Jones time series. The timeline is composed of discrete monthly data points. *Source* Author

Table 8.1 Performance measures for the Dow Jones dataset

τ	Baseline	Precision	Recall
1	6	94	94
2	12	78	88
3	17	73	81
4	23	67	74
5	28	60	72
6	33	57	62

The baseline probability for prediction is $p(SIGSLOP_{i+\tau} = -1/SIGLOP_i = 1)$. All results are given as round percentages

I used a machine-learning classifier to predict a trend reversal. To secure the validity of the findings, I used a procedure known as *ten-fold cross-validation*, where the classifier uses some data points to learn a pattern and test itself on other cases. This approach is better than fitting a model to the data because it limits the possibility of over-fitting.

The performance measures of the classifier in predicting the transition are presented in Table 8.1. The measures are Precision (Positive Predictive Value) and Recall (True Positive Rate). The performance measures concern the prediction of SIGSLOP = −1.

The baseline for prediction is obtained by measuring the probability of observing a trend reversal given the exponential growth in the previous window. The results show a significant improvement in prediction over the baseline. For example, given a time window showing exponential growth in the stock price, the probability of observing trend reversal in the next time window (i.e., W_{i+1}) is extremely low (6% or $p = 0.06$). The reason is simple. Shifting our window of attention one month ahead gives an overlap of 11 observations between the first and the second time frames. The first and the second time frames overlap in most of their months, so a trend reversal resulting from a single month is highly unlikely. How unlikely? Only 6% of the cases show trend reversal.

Changing a trend does not mean that the market is collapsing. It just means that the strong exponential growth is changing. It can change into weak exponential growth, non-growth, or a decline. In this context and assuming full ignorance except for this conditional probability, we should predict only 6% of trend reversals. However, using ASY and StS in the predictive model, we gain 94% precision and 94% recall. This means that the model successfully predicted 94% of trend reversals, and when predicting this changing dynamic, it was precise in 94% of the cases. Using this model, we may improve the identification of a changing trend by 88%.

These results show a significant improvement in prediction over the baseline. Impressive as it may be, the relevance of these findings may not be totally clear to the reader. In the next section, I explain its relevance for understanding the idea of betting against the crowd.

Betting Against the Herd

Herding may be expressed by a bullish market, where the prices are pushed upward by the enthusiastic crowd, self-enhancing its excitement through a positive feedback loop and the exponential spread of information. In this context, betting against the crowd does not mean avoiding the party. It just means leaving on time, before getting too drunk and hitting a wall. The crowd itself may continue its trajectory through inertia. As we learned from Newton, an object in motion will remain in motion in a straight line unless acted upon by external forces. Crowds may move forward like a herd until acted upon by the force of collapse. Understanding that this herding behavior is constrained by limited energy resources may be important for finding out when to leave the party.

A more general lesson I have emphasized throughout the book is the importance of gaining a profound understanding, by which I mean understanding basic and foundational processes underlying the crowd's behavior. As an individual operating on the margin of the madding crowd, we may have the advantage of distancing ourselves from the crowd through profound simplicity combining deep understanding, simple signals, and the idea of being non-stupid.

In sum, this chapter is another instance of betting against the crowd. I have suggested using two simple signals built upon a deep understanding of market dynamics and foundational scientific ideas. Simple signals are not always available, and leaving the crowd is not always possible. However, the book does not claim to provide a manual for the wise trader, and this chapter is illustrative only. In particular, it illustrates the possibility of distancing yourself from the madding crowd and it shows that distancing yourself from the crowd may have enormous advantages. Again, I pointed to the importance of timing and context, recognizing the relevant time frame and simple signs indicative of deep underlying processes. The general lessons one may learn from my examples are far more important than the technical aspects, because the latter are context-dependent and may vary significantly from one case to another, while the general lessons and the right attitude are far more important, and these may be carried across contexts.

References

Bandt, C. (2020). Order patterns, their variation and change points in financial time series and Brownian motion. *Statistical Papers, 61*, 1565–1588.

Cohen, J. (1988). *Statistical power analysis for the behavioral sciences* (2nd ed.). Lawrence Erlbaum.

Emmert, H. R. (2018). Selling out: The financialization of contemporary art. In *Proceedings of the 11th International RAIS Conference on Social Sciences* (pp. 264–269).

Kameda, T., & Hastie, R. (2015). Herd behavior. In R. Scott & S. Kosslyn (Eds.), *Emerging trends in the social and behavioural sciences* (pp. 1–14). Wiley.

Litimi, H. (2017). Herd behavior in the French stock market. *Review of Accounting and Finance, 16*(4), 497–515.

McNamee, L., & Ledley, F. (2012). Patterns of technological innovation in biotech. *Nature Biotechnology, 30*, 937–943.

Neuman, Y., & Cohen, Y. (2022). Predicting change in emotion through ordinal patterns and simple symbolic expressions. *Mathematics, 10*(13), 2253.

Neuman, Y., & Cohen, Y. (2023). Unveiling herd behavior in financial markets. *Journal of Statistical Mechanics: Theory and Experiment, 2023*(8), 083407.

Pastor-Satorras, R., Castellano, C., Van Mieghem, P., & Vespignani. (2015). Epidemic processes in complex networks. *Reviews of Modern Physics, 87*, 925.

Raafat, R. M., Chater, N., & Frith, C. (2009). Herding in humans. *Trends in Cognitive Sciences, 13*, 420–428.

West, G. (2018). *Scale: The universal laws of life, growth, and death in organisms, cities, and companies*. Penguin.

Xiong, J., Liu, Q., & Zhao, L. (2020). A new method to verify Bitcoin bubbles: Based on the production cost. *The North American Journal of Economics and Finance, 51*, 101095.

Yook, S.-H., & Kim, Y. (2008). Herd behavior in weight-driven information spreading models for financial market. *Physica A: Statistical Mechanics and Its Applications, 387*, 6605–6612.

Zhang, R., Yang, X., Li, N., & Khan, M. A. (2021). Herd behavior in venture capital market: Evidence from China. *Mathematics, 9*(13), 1509.

9

Unraveling the Complexities of Chronic Armed Conflicts: Patterns, Predictability, and Uncertainties

Can We Learn from the Past? And How Much?

> History teaches us nothing. (Attributed to Hegel)

In this book, I repeatedly question whether we can discover informative patterns that may improve our position as individuals in and against the crowd. One common opinion is that history teaches us nothing, as the German philosopher Hegel allegedly said. According to this opinion, each new situation we must address will have a unique, singular, and contextual nature. The use of general and abstract patterns from the past will be largely irrelevant to the way we should act now.

Think, for example, about the fall of the great empire of Rome. Numerous books have been written about the rise and fall of the Roman Empire, but is there a lesson we may learn and apply to the current American empire? In this view, general lessons are too general to be used as guidelines for the best way to act, and particular guidelines for preserving the current American empire will not be trivially transmitted from an old empire to modern and postmodern empires. So, history seems to teach us nothing.

An opposing opinion is that we cannot learn without remembering the past and identifying trends and patterns that can be used for our present and future conduct. But both opinions are both right and wrong when taken out of context. It is highly important to identify trends, but at the same time, it is important to avoid fixing on an idea while ignoring the uncertainty resulting from the particularity of present and future events.

Y. Neuman, *Betting Against the Crowd*, https://doi.org/10.1007/978-3-031-52019-8_9

This is a very important point when dealing with crowds and betting against them. As a result of the interaction between members of the crowd, constraints may accommodate and limit the crowd's degrees of freedom and its ability to respond flexibly and promptly to changing circumstances. This is what we found when we analyzed the behavior of football teams (Neuman & Vilenchik, 2019). The crowd may, therefore, be prone to the fixation problem and find itself blindly following a trend. There is a comforting pleasure in holding on to simple and well-known explanations. It reduces uncertainty. However, like all slaves to a fixed idea, we may behave like the stupid crocodile mentioned in Peter Pan. For the individual seeking to bet against the crowd, an opportunity may be created by adopting a complex view of the situation that combines the general understanding of trends and the uncertainty built into the situation. In this chapter, I would like to inquire about this approach in the painful context of a chronic and long-term armed conflict.

The Long Memory of a Conflict

Long-armed conflicts, such as the one between the Israelis and the Palestinians, seem to express pattern-like behavior that may result in a feeling of déjà vu. Whenever I personally experience another round of the conflict, I ask myself whether we have not already experienced the same event. This feeling may be explained by patterns that involve long-term memory, cycles of retaliation, and anti-persistent behavior where, in the short run, an increase in fatalities is followed by a decrease and vice versa. By observing the cycle of violence, we may notice simple and repeating patterns. We can always recall a previous case of escalating violence whenever we observe a new one. If a wave of terror is repressed, we may recall a previous case where such a wave was repressed, only to raise its ugly head again shortly afterwards. The feeling of déjà vu is a clear expression of repeating patterns and trends, which are temporal patterns.

To better understand a conflict, we may first examine its most concrete aspect: the number of lost lives. Power law distributions are among the most widely discussed patterns associated with armed conflicts. It has long been argued that fatalities in armed conflicts, specifically wars, follow power law distributions (Richardson, 1948), although some have questioned this hypothesis (Zwetsloot, 2018). However, the power law does indeed seem to describe this aspect of war. A minority of wars are responsible for a disproportionate number of fatalities, while many small conflicts make their own marginal and devilish contribution to the death toll.

For identifying the distribution of fatalities in the Israeli context, I used 8372 data points[1] representing the number of Israeli and Palestinian fatalities since 2000. Figure 9.1 presents the percentage of fatalities with 95% confidence interval error bars.

One can easily see a power law where the number of fatalities is 0 or very low in most cases, and a long tail stretches to 341 fatalities per day. A linear regression model fitted to the log transformed timeline and number of fatalities produced a regression line with negative slope ($B = -0.77$), indicative of a power distribution. Such a distribution is characterized by a very long and theoretically unbounded tail. This is a highly important point. This chapter was written before the recent attack on Israel. As this chapter was being completed, the maximum number of fatalities broke a new record, emphasizing in real-time the long tail's unbounded nature. It seems that in real life, it can always be much worse than we have ever experienced or imagined, and the probability of experiencing this "worse than we know" event is higher than we usually believe, specifically under the wrong assumption of a "Normal" distribution.

The long tail of fatalities characterizing armed conflicts is a matter of fact and has profound implications for understanding, predicting, and managing

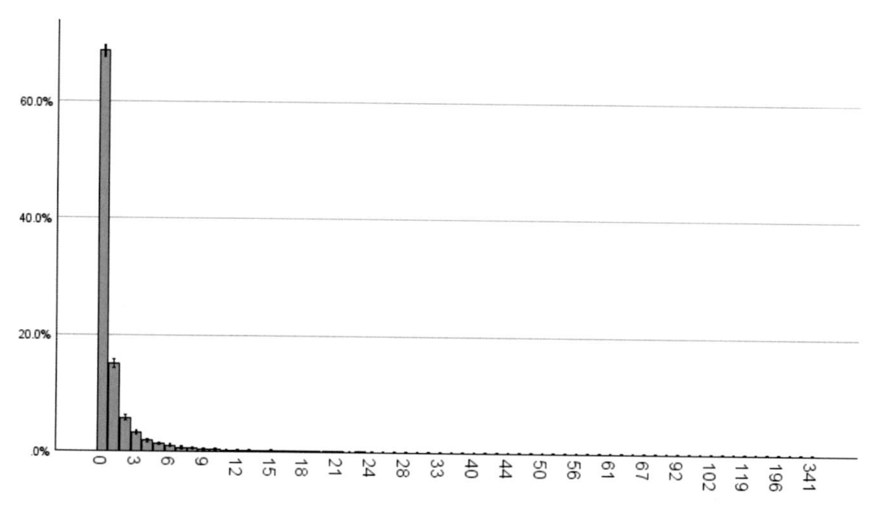

Fig. 9.1 Percentage of fatalities with 95% CI error bar. *Source* Author

[1] https://statistics.btselem.org/en/all-fatalities/by-date-of-incident?section=overall&tab=overview. For the analysis, we merged data from several data sets.

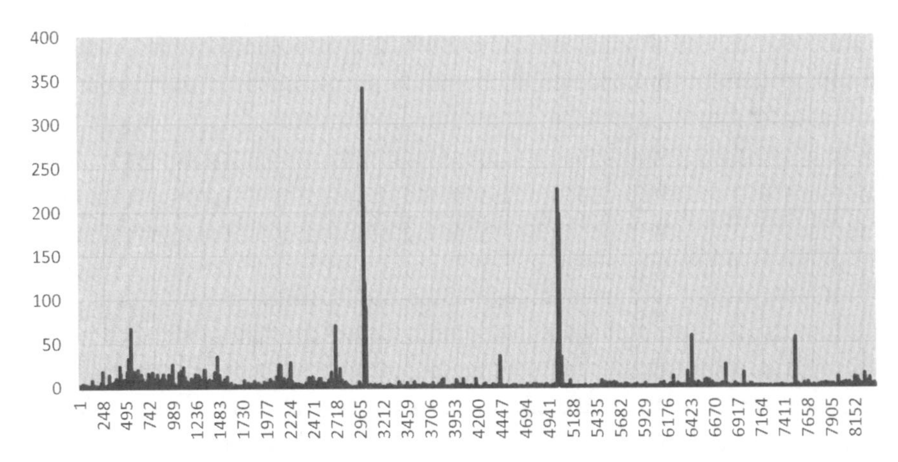

Fig. 9.2 Daily count of fatalities in 8156 data points. *Source* Author

these conflicts. One clear implication of the power law distribution is the unpredictability of rare and extreme events, despite the feeling of déjà vu grounded in the conflict's identified patterns. Figure 9.2 presents the daily number of fatalities, where noticeable peaks express the long tail of fatalities.

These extreme events are *mostly unpredictable*, as illustrated by the recent attack on Israel and the unprecedented fatalities among Israelis. In this context, an important and basic scientific question is: why are rare and extreme events, such as wars, characterized by a low predictive signature?

The nonexistence of a strong predictive signature may be counterintuitive to some decision-makers, because we might naively expect the arrival of extreme events that naturally require a significant amount of energy to be preceded by a strong signature. In other words, a decision-maker, such as the secretary of defense, can ask his advisors why an event such as a war is unpredictable despite the enormous effort required to launch a war in terms of energy.

The answer is that some wars can be predicted quite specifically, shortly before they erupt. Intelligence agencies identified the preparation of the Russian army to invade Ukraine. However, how many people could have predicted the war in Ukraine months ahead? In some contexts, such as the recent Israeli context, the exact timing of the war was unpredictable, given the enemy's efforts to hide its plan, the surprise attack, and the failure of intelligence to identify early warning signals or to recognize warning signals as such. However, a more general and scientifically grounded answer to the question (West, 2016) is that rare and extreme events have a strong component of nonlinear, multiplicative interactions, similar to the mechanism hypothesized to result in freak or rogue waves described in Chap. 6. This non-linear dynamic results in rare, unexpected, and extreme events.

If the above answer is scientifically grounded, then (1) we can explain why extreme events cannot be successfully predicted *despite* identified trends and patterns, and (2) we can hypothesize that events with a high number of fatalities should be preceded by a period where the distribution of fatalities indicates high complexity/entropy/uncertainty.

Let me explain these hypotheses. First, we may extrapolate to the near future when we observe a clear trend in fatalities. A trend indicates that the system has a long memory. A system with a long memory or long-range dependence is a system where the value of a measurable variable is correlated with past measurements. This correlation is indicative of a trend, but it may blind us with respect to prediction. I shall explain and justify this point later.

Second, one potential source of failure in predicting an extreme event is *our ignorance of our ignorance*. Recall the football example. There are three main components to our analysis and understanding. We have Athena, who generates the approximated trend. Our positively oriented wisdom directs us toward identifying patterns and measuring trends. We also have Lady Fortuna, who reminds us that we must take uncertainty into account, and Hermes, who reminds us that enemy tricksters would try to hide their malevolent plans. As trends are not enough for prediction, we must measure the uncertainty and complexity of the situation. This may add value to our prediction efforts, as will be seen later.

I analyzed 8372 data points from 29 September 2000 to 6 June 2022. The timeline has been segmented into 266 nonoverlapping 30-day windows. These time windows are abbreviated as TW. I also measured the number of Israeli and Palestinian fatalities in each TW.

First, I asked whether this number could be predicted from the number of fatalities in the previous TW. The idea is that conflicts may be characterized by short and long-range dependence, indicating a trend nurtured by positive escalation or de-escalation formed through feedback loops.

The correlations between the number of fatalities in a TW and those in the 12 lagged TWs are presented in the next graph. The correlations were computed using Spearman's rank correlation coefficient. All results were found to be statistically significant at $p < 0.001$. An exponential regression analysis was conducted to model the relationship between the time lag (1–12) and the correlation between the number of fatalities in a target TW and the number of fatalities in the lagged TW. In other words, I tested the correlation between the number of fatalities in a TW and the number of fatalities in a previous time window and analyzed these correlations as a function of the lag, or how distant the past measurement was from the present one. Figure 9.3 presents this relation (adjusted $R^2 = 0.691$, $p < 0.001$, $B = -0.048$).

Given that the slope is negative, it implies exponential decay. Specifically, a negative value of B indicates that the autocorrelation decreases as the time lag between observations increases. In this case, the value $B = -0.048$ suggests a relatively *slow* decay in correlation over time. These findings may indicate a long-range dependence characterized by a *slower-than-exponential decay rate*. Therefore, it is indicative of long-range memory. Regardless of the system's long-term memory and trend-like behavior, it is not enough for prediction. Let me explain these findings using a simple example.

Imagine that I measure the correlation between the words you produce. You share the happy news that you have just proposed marriage to Rachel, your sweetheart. You say: "I love ..." and before you complete the third word, I can guess it is "her" or "Rachel." There is a context to the conversation involving correlation between words. Guessing the next few words may be relatively easy, but can I guess the word you will generate one minute from now? Two minutes from now or even four minutes from now? As time goes by, we expect the correlation to decay exponentially. However, if the decay is slower than exponential, we can conclude that the measurements along your timeline, whether words or fatalities, are correlated. This means that

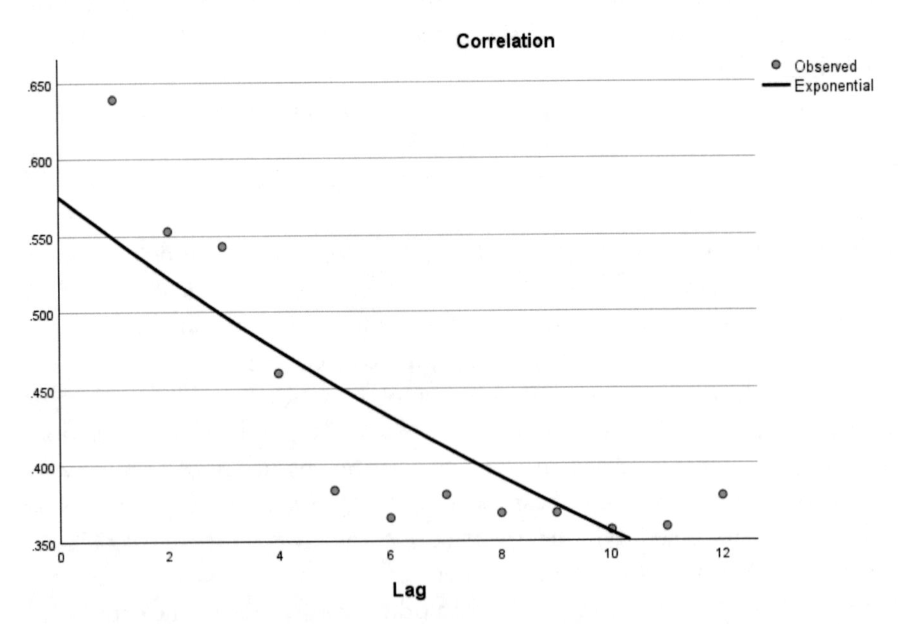

Fig. 9.3 Exponential model fit for lag and the Spearman correlation coefficient between the fatalities in a TW and those in a lagged TW. *Source* Author

the system presents some form of long-term memory. Such a memory is characterized by a long-term dependence between the measured values.

In the above analysis, I found an indication that a long-term memory exists for the timeline of fatalities. This is a trivial finding assuming an upward or a downward trend. Measure a persistent bull market or a persistent bear market, with a decrease in stock values, and you will find long-term memory. The existence of long-term memory seems promising as it may help us to reduce our uncertainty about future events. So, it seems that Hegel was wrong. Identifying a trend in the number of fatalities, we can look at the past and use it to predict the future. But not so fast, as I explain in the next section.

Order Is Never Enough

I found a positive correlation between the number of fatalities in a given TW and those in the previous TW ($r = 0.669$, $p < 0.001$). Given this correlation, one may naively hypothesize that the prediction of fatalities is a simple challenge addressed by persistent trends and the long-range memory of the system. However, Spearman's rho correlation has no *predictive* meaning. Moreover, it cannot provide a clear measure of explanation, such as the "explained variance." To illustrate this point, I tested several regression models with the number of fatalities in a TW (i.e., the target TW) as the dependent variable and the number of fatalities in the previous TW as the independent predictive variable.

The model that gained the highest explained variance was a power regression model with an adjusted R^2 of 0.405. This means it explained approximately 40% of the variance in fatalities but left 60% unexplained. We can analyze the error in the model in an attempt to understand the meaning of this explained variance for predicting the *number* of fatalities. As Jansen et al. (2022) show, numerical data about people is repeatedly wrong. Trying to use a correlation for prediction or even a simple regression model may result in large errors while dealing with the *number* of fatalities in long-tail distributions. For example, the average error in predicting TW fatalities is $N = 22$ (95% CI 6–37), with a huge range of errors: $R = 1607$. It means that even the best regression model *fitted* to the data cannot avoid huge predictive errors resulting from extreme events such as wars. The positive correlation seems significant only in a specific statistical sense, and the best regression model presents a large unexplained variance, which is translated into a range of errors that a reasonable decision-maker cannot accept.

How can we explain this gap between the existence of trends in the data and the acute failure in prediction? One possible hypothesis is that some extreme and rare events that heavily influence our prediction have a strong component of non-linear, multiplicative interactions that converge to a rare and powerful outcome. This is the same dynamics that I recognized while analyzing Hungarian political parties. Some hidden and underlying currents may suddenly converge to generate a freak wave of terror. Therefore, I hypothesized that measuring this component of uncertainty in the system might *improve* our prediction over the one obtained from a trend in fatalities. To measure this component, I used the Tsallis entropy explained previously and used (Tsallis, 2014; Tsallis et al., 1998, 2003).

Tsallis Entropy, Again

I used Tsallis's version of entropy when analyzing the nationalism level of political parties in Hungary. Here, I use it again, but first, let me remind the reader why I use this specific measure of uncertainty. The Shannon information entropy measure is a natural choice for measuring a system's unpredictability. The Tsallis entropy extends this concept to account for systems with power-law behavior or long-range correlations. The Tsallis entropy is typically used when the underlying system exhibits non-extensive and non-additive behavior. Traditional entropy measures, such as the Shannon entropy, assume that the interactions between system components are *short-range* and *additive*. However, in systems with long-range interactions, such as the system we are analyzing in this case, the Tsallis entropy may provide a more appropriate description by accounting for the non-additive and non-extensive nature of the interactions. As we scale up in size, the expected amount of disorder does not scale linearly (this is non-extensivity) but produces a more disordered and unpredictable whole than its smaller version. The same is true for non-additivity, meaning that the entropy of the composite system is not the sum of the entropies of its parts.

In this context, the entropy index q included in the expression for the Tsallis entropy plays an essential role. The parameter q used to compute the Tsallis entropy introduces non-additivity and non-extensivity. A $q > 1$ case may better represent non-additive and non-extensive systems (Tsallis & Tirnakli, 2010). This implies that the joint system exhibits a higher degree of disorder or complexity than the sum of its parts and does not scale linearly with size. The Tsallis entropy is particularly relevant in cases where the underlying probability distribution exhibits heavy tails, a characteristic of power-law distributions. When q is chosen appropriately, the Tsallis entropy can capture the statistical properties of systems governed by power-law distributions. In other words, q serves as a parameter that allows us to tailor the Tsallis entropy to different types of distributions, including those with long tails. When q is adjusted to reflect the underlying distribution, the Tsallis entropy provides a more accurate measure of entropy for systems that follow power-law behavior.

Discretizing the Dataset

To measure the Tsallis entropy in each TW, I first discretized the fatality data in each TW by converting each TW into a discrete distribution of fatalities. To do this, I observed the distribution of fatalities in each TW and put them into a well-defined number of "boxes" or "bins" to ease data analysis. The challenge is in deciding the number of boxes and their width.

I adopted a cognitively motivated approach to generate a distribution that aligns with human perception and may be more useful to the decision-maker's logic. First, I relied on the idea previously presented that the human mind is logarithmically grounded (Varshney & Sun, 2013) and differentiates between categories of perception using the just noticeable difference (JND). The JND, or difference threshold, is a concept in psychology and sensory perception that refers to the smallest detectable difference between two stimuli. It is the minimum amount of change in a stimulus that can be perceived by a person.

The JND is fundamental in fields like psychophysics, which studies the relationship between physical stimuli and the sensations and perceptions they evoke. The German psychologist Ernst Weber developed the concept in the early nineteenth century, an idea that was later refined by his colleague Gustav Fechner.

Weber's law, which is closely related to the concept of the JND, states that the JND is proportional to the magnitude of the original stimulus. In other words, the JND is not an absolute value but a relative one. For example, if you were holding a weight in your hand, you would notice a smaller additional weight if the original weight was light rather than heavy. This law of perception corresponds well to our commonsense experience. For a poor and hungry person, a piece of bread and one dollar is not only a noticeable difference but a significant noticeable difference and a source of happiness, while one additional dollar in the pocket of a rich person would go unnoticed.

Table 9.1 The five classes of fatalities

Class	Fatalities	Fatalities range	Percent fatalities	Months in class	Percent of months
1	116	0–3	1.5	56	21
2	339	4–9	4.4	58	21
3	954	10–25	12.5	62	23
4	2186	26–63	28.6	52	20
5	4033	64–1417	52.9	38	14

Column 3 presents the range of fatalities included in the class, column 5 presents the number of months included in each class, and column 6 presents the (rounded) percentage of these months

To form categories and transform the fatalities into a discrete distribution, I computed the relative JND (RelJND):

$$\mathrm{RelJND} = \frac{\sigma}{R} * 2$$

and the binned score of fatalities as

$$\mathrm{RND}(\mathrm{Log}_{10}(\mathrm{Fatalities})/\mathrm{RelJND})$$

To differentiate more clearly between the categories, I used JND * 2, and merged two categories with an extremely small number of data points. The final procedure produced five different classes of fatalities that correspond to the logarithmic nature of the human mind and the idea of the JND. The five classes are presented in Table 9.1.

We see that class 1 includes 116 fatalities who died in 56 TW. Regarding class 5, which represents the highest class of fatalities, it is important to note that 14% of the TW are responsible for 53% of the fatalities in the dataset. This category corresponds roughly to frequently rare events (Shyalika et al., 2023).

Predicting the Class of Fatalities

First, I tested the relevance of the q value for the entropy measurements. Determining the exact value of q is an open challenge. As in my previous work, I chose three q values: q = 0.2, 1 (Shannon entropy), and 2.

I specifically focused on performance in predicting class 5. First, I used a neural network with two predictors: the number of fatalities in the previous time window and the Tsallis entropy score of the fatality distribution in the previous window. I performed seven runs of the analysis with 70% of the dataset for training and the rest for tests. In terms of the true classification rate (TCR) (i.e., recall), the model with $q = 2$ outperformed the models with $q = 1$ and $q = 0.2$ (60% vs. 41% and 47%, respectively). Using the model, I correctly identified in advance, and with 60% success, which class of fatalities I would see in the next TW. Therefore, I found it justified to measure the entropy of fatalities using a Tsallis entropy with $q = 2$.

I hypothesized that the entropy score of a TW preceding a TW with high fatalities (e.g., class 5) should be higher than the entropy of one preceding a window with low fatalities. To test the hypothesis, I first looked to see whether there was a significant difference between the entropies of the different classes. I used the Kruskal–Wallis H test with the class as a grouping variable and the six lagged scores of Tsallis entropy as the test variables. In other words, I compared the Tsallis score of the six lagged TWs (− 1 to − 6) preceding a target TW across the five classes. Figure 9.4 presents the entropy score (lag 1) for each fatality class.

The Kruskal–Wallis H test was statistically significant ($H = 113.88$, $p < 0.001$), but the mean rank for each class is more important. The mean rank for each class and lag 1 for the Tsallis entropy is presented in Table 9.2.

We see here a clear difference in the entropy levels of the different categories of fatalities. The results support the hypothesis that TWs characterized by high fatalities are preceded by TWs with a higher level of entropy. However, the first analysis measured *differences* only. Next, I tested the

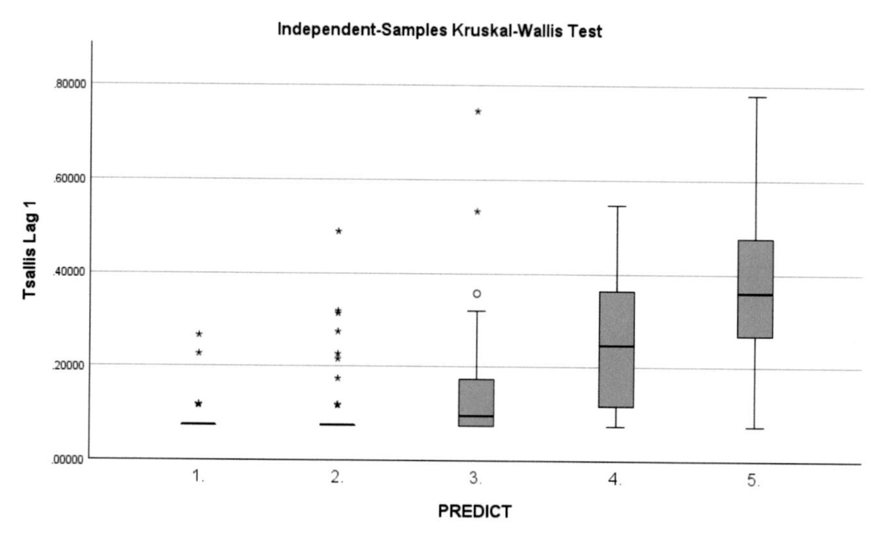

Fig. 9.4 Entropy score (lag 1) for each fatality class. *Source* Author

Table 9.2 The mean rank of the Tsallis entropy (lag 1) for the five classes of fatalities

Class	N	Mean rank
1	56	87.26
2	58	91.97
3	62	123.24
4	52	182.79
5	37	212.93

hypothesis that the entropy measurement of fatalities may be used to better predict the class of fatalities in the target TW.

In analysis 2, I used multinomial logistic regression analysis and the naïve Bayes classifier. First, I used multinomial logistic regression with class 5 as the reference category. The predictors were the six lagged measurements of fatalities. Next, I tested the model with the fatalities and six lagged entropy measurements. The results are presented in Table 9.3 with the TPR performance measure (i.e., correct classification). All percentages are rounded.

We can see that the combined model outperformed the model with the fatalities alone, specifically with respect to class 5, which is the focus of our interest. The combined model correctly identified 62% of the TWs in which the number of fatalities belongs to class 5. This is a significant improvement in prediction compared to the prediction relying on the baseline (i.e., 14%) and the one obtained from the trend in fatalities alone. Using the naïve Bayes classifier, we get the results shown in Table 9.4.

Table 9.3 Multinomial logistic regression performance for the two models

Class	Baseline	Fatalities only	Fatalities + entropy
1	22	70	60
2	22	14	33
3	23	32	39
4	20	28	52
5	14	24	62

Table 9.4 Naïve Bayes performance in classifying classes of fatalities

Class	Baseline	Fatalities only	Fatalities + entropy
1	22	13	86
2	22	9	19
3	23	98	48
4	20	6	46
5	14	30	70

The naïve Bayes model selected only six predictors: four lagged measurements of fatalities and two lagged entropy measurements. On average, it performed better with the combined model than multinomial logistic regression. In total, 70% of the TWs in which the number of fatalities belongs to class 5 were identified, although the percentage of this class in the dataset is only 14%. Therefore, this supports the hypothesis that entropy improves predictions over those obtained from the fatalities trend.

Predicting Surprise

Next, I defined a new measure called Surprise: $\text{fatalities}_n - \text{fatalities}_{n-1}$. The distribution of Surprise is presented in Fig. 9.5, where a positive surprise implies a higher number of fatalities in the present TW than in the previous TW.

To test the ability to predict the *absolute* level of surprise, I used three main variables: Fatalities, Entropy, and the Relative Entropy measure (KL divergence). The KL divergence quantifies our level of surprise when observing distribution P using distribution Q. For the dataset, I defined distribution P as the distribution of fatalities at TW_n and Q as the distribution of fatalities at TW_{n-1}. There was no overlap between the TW I tried to predict and the TW where I measured the KL divergence.

In addition, I computed the difference between fatalities at lag n and lag $n - 1$, for lags $i = 1$ to 12. The sum of this difference was called "Trend."

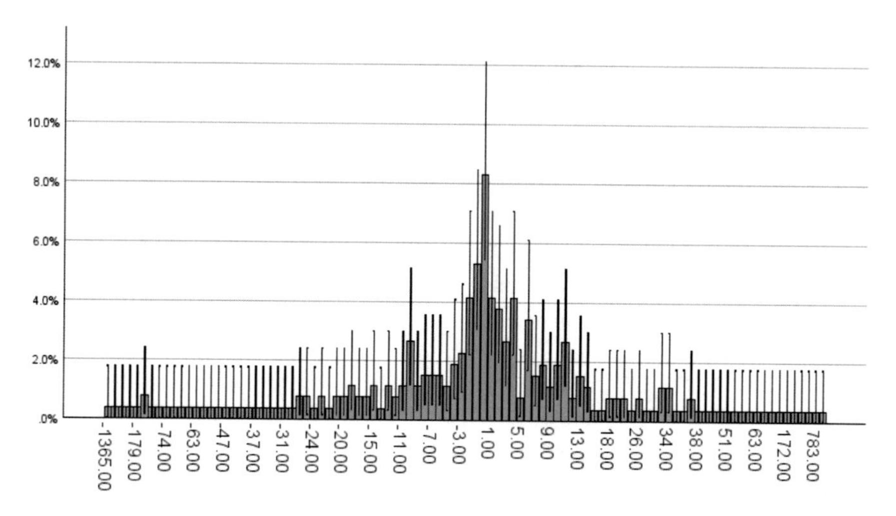

Fig. 9.5 The distribution of surprise. *Source* Author

Table 9.5 Performance measures for predicting surprise

Class	Baseline	Avg. absolute surprise	True positive rate (recall)
1	18	1	60
2	19	2.9	26
3	24	8.17	62
4	20	19.26	34
5	20	173.88	66

Trend was used to determine whether there was an upward or downward trend in the fatalities. I used the three lagged scores of Fatalities, Entropy, and KL divergence, and the single score of Trend. For prediction, I discretized the absolute surprise score using the same procedure of discretization described before. Table 9.5 presents the percentage of each class. This percentage was used as a baseline for estimating the classification performance. In addition, the table presents the average surprise score of each class and the TPR as generated by multinomial logistic regression.

As can be seen, the model improved prediction over the baseline of surprise. This means that the *absolute* level of surprise can be well predicted using measures of trend and entropy.

The absolute size of the surprise is not as interesting as its direction. Therefore, I discretized the raw surprise score using the same approach as before. This procedure resulted in seven surprise classes. Next, I used the naïve Bayes classifier with the following class variables: The six lagged entropy measurements, KL divergence, and fatalities + Trend and a measurement of the Hurst exponent performed for the year preceding the target time window. The Hurst exponent tests whether a time series is chaotic, trend-like, or presents an anti-persistent trend when an increase is followed by a decrease or vice versa. The classifier chose ten features for the final model: two entropy scores, four KL scores, the Hurst exponent, and Trend. Table 9.6 presents the minimum and maximum surprise for each class and the classifier's performance compared to the baseline.

These results are particularly interesting regarding the highest level of surprise, in which classification has been improved by 36% over the baseline. The highest level of surprise (class 6) involves the unpleasant surprise of much larger fatalities than the previous TW. Improving the prediction by over one-third may be important for various practical contexts.

Table 9.6 The naïve Bayes performance in classifying surprise

Class	Baseline	Min	Max	TPR
0	9	0	0	35
1	18	− 1365	−15	71
2	13	− 14	−5	32
3	18	− 4	1	30
4	18	2	8	73
5	13	9	26	40
6	11	27	1384	47

Crowds, Conflicts, and Lady Fortuna

The aim in this chapter was to enrich our understanding of armed conflicts by carefully analyzing a single case. The trend expressed by the number of fatalities along a timeline is a good predictor of future fatalities in a system that expresses a long-range memory. Long-range dependence suggests some level of persistence or memory in the system. This implies that past events continue to influence present and near-future events, indicating a degree of predictability. My main argument is that the prediction of fatalities may be enhanced, specifically in the case of extreme events, by analyzing the entropy of the time windows preceding our near future. Higher entropy indicates higher unpredictability and sensitivity to rare events, implying that extreme events or outliers impact the system more. However, there seems to be an incoherence between the persistent and predictable behavior implied by a long-range memory and the predictive value proposed by the entropy measure.

The coexistence of long-range dependence and higher Tsallis entropy can be understood in the following way. First, an armed conflict is a complex adaptive system. In such systems, predictable patterns can coexist on certain time scales (long-range dependence) and unpredictable, chaotic behavior on other time scales sensitive to *long-range interactions*. Second, as realized by Bateson long ago (Harries-Jones, 1995), social systems operate across multiple scales of time and space. On a longer time scale (captured by long-range dependence), underlying structural or systemic factors may drive identifiable patterns or cycles. In contrast, on a shorter time scale (captured by higher Tsallis entropy), the system may be influenced by unpredictable events that may spark a serious escalation in the short run.

In summary, the coexistence of long-range dependence and higher Tsallis entropy may indicate the complex and adaptive nature of an armed conflict. Understanding and modeling such systems requires us to consider dynamics across multiple time scales and acknowledge that predictable and unpredictable elements contribute to the system's overall behavior. This understanding must be considered by decision-makers who may be blinded by their ignorance of complexity and reliance on simple statistics. If there is a single lesson to be learned from the current chapter, then it is this: *Do not trust trends!* Trends are like mythological sirens that tempt poor sailors with sweet voices. Do not trust them. Use them, but do not trust them.

What lesson can we learn about crowds and the empowered individual? The lesson is painful in its simplicity. Despite the trend-like behavior of crowds and our ability to discern certain patterns, this understanding is deeply flawed when we try to identify the divergence of the crowd from normative behavior, if there is such. In this context, betting against the crowd is betting against the crowd that observes a crowd and fools itself into believing that it is Lady Athena in her full glory. The empowered individual, aware of the existence of Lady Fortuna and Mr. Hermes, may suspect the crowd's biased understanding and seek signs indicating a different direction.

As the legendary magician Harry Houdini explained, knowledge is not only power. Knowledge is safety. Knowing that you do not know and must include the uncertainty in your worldview may be a safety valve against the blindness and fixation of the crowd.

References

Harries-Jones, P. (1995). *A recursive vision: Ecological understanding and Gregory Bateson.* University of Toronto Press.

Jansen, B. J., Salminen, J., Jung, S. G., & Almerekhi, H. (2022). The illusion of data validity: Why numbers about people are likely wrong. *Data and Information Management, 6*(4), 100020.

Neuman, Y., & Vilenchik, D. (2019). Modeling small systems through the relative entropy lattice. *IEEE Access, 7*, 43591–43597.

Richardson, L. F. (1948). Variation of the frequency of fatal quarrels with magnitude. *Journal of the American Statistical Association, 43*(244), 523–546.

Shyalika, C., Wickramarachchi, R., & Sheth, A. (2023). *A comprehensive survey on rare event prediction.* arXiv preprint arXiv:2309.11356

Tsallis, C. (2014). An introduction to nonadditive entropies and a thermostatistical approach to inanimate and living matter. *Contemporary Physics, 55*(3), 179–197.

Tsallis, C., & Tirnakli, U. (2010, December). Nonadditive entropy and nonextensive statistical mechanics—Some central concepts and recent applications. *Journal of Physics: Conference Series, 201*(1), 012001.

Tsallis, C., et al. (1998). The role of constraints within generalized nonextensive statistics. *Physica A: Statistical Mechanics and Its Applications, 261*, 534–554.

Tsallis, C., et al. (2003). Introduction to nonextensive statistical mechanics and thermodynamics. arXiv preprint cond-mat/0309093.

Varshney, L. R., & Sun, J. Z. (2013). Why do we perceive logarithmically? *Significance, 10*(1), 28–31.

West, B. J. (2016). *Simplifying complexity: Life is uncertain, unfair and unequal*. Bentham Science Publishers.

Zwetsloot, R. (2018). Testing Richardson's Law: A (cautionary) note on power laws in violence data. Available at SSRN 3112804.

10

Mentsh Trakht un Got Lakht: Final Lessons in Individuality and Collective Dynamics

The opening scene of the film "Andrei Rublev" (1966) by Andrei Tarkovsky takes place in Medieval Russia. A man is trying to launch a hot-air balloon. This is an improvised and somewhat rough-and-ready version of the hot-air balloons we know today. The man and his small group of friends are in a hurry. A furious mob is arriving from the river, and we understand that their aim is to destroy the venture. Finally, the balloon takes off, and the man who is attached to it by simple ropes is excited. He is flying! Free as a bird, he travels a short distance across the sky, before crashing to the ground. The mob may well then have destroyed the balloon.

Immersed in symbolism, this opening scene confronts the clash between a free spirit and the madding crowd. There is an immanent conflict between the crowd and the individual, at least in cultures where the individual plays a significant role. One of these cultures is ancient Judaism. Judaism produced a corpus of texts known as the Mishna. The Mishna Sanhedrin is a significant text that deals with various aspects of the legal system, including the establishment and functioning of courts. It's part of the Mishna, a foundational work of Jewish oral law. We don't know exactly when it was written, but some date it to 200 AD. In the Mishna (Neusner, 1974, 4:5), we may read:

> The greatness of the Holy One, blessed be He, is that a person stamps many coins with one seal, and they all resemble each other, but the King of Kings, the Holy One, blessed be He, stamps each person with the seal of Adam the First, and none of them is similar to his fellow. Therefore, each of us must say, 'For my sake, the world was created.' (Author's translation from the Hebrew version)

Y. Neuman, *Betting Against the Crowd*, https://doi.org/10.1007/978-3-031-52019-8_10

This is an interesting idea. Old coins were stamped with the king's image. Each and every coin represented the king, who stamped them, and all coins looked *the same*. In contrast, God, metaphorically described as the king of all kings, "stamped" each person with the seal of Adam, the first person. However, none of us is the same as the others. *We are all unique*. This uniqueness has deep implications for our moral conduct. For the ancient Jewish thinkers, and in contrast with some of their fake modern heirs, each human being is unique and deserves full respect as if the whole world was created for him or her, as if he or she was the first man or woman on Earth.

In practice, we know that this ancient imperative is repeatedly violated. We are all unique but never alone; and when we are never alone, the crowd seeks to destroy our uniqueness. In Andrei Rublev, the crowd literally murders the individual who is seeking to fly. Conformity to the crowd with its duplicated copies of the king seems to be the violent crowd's first imperative. The conflict between the crowd and the individual is unresolved for those emphasizing individuality, because "we are all unique but never alone." We live as a part of larger circles of other human beings, and the formation of crowds is a reality that sometimes conflicts with our individuality.

In this book, I have tried to delve deeper into the collective mind to understand better how the individual may find his or her place within and against the crowd. My attempt has been deeply influenced and biased by my personal experience. Experiencing the attempts to destroy democracy and the surprising and vicious attack of a mob from Gaza cannot go unnoticed, and they cannot go without leaving their mark. The lessons learned in this book are trivial in a deep sense, and I am proud of it. Reaching a trivial conclusion, albeit from a scientific perspective, has the comfort of appealing to common sense, which is not so common.

Despite their allegedly simple appearance, I have explained that crowds are the emerging product of human interactions. They combine order and chaos, logic and trickery. When we study them, we must be cautious, modest, reflective, and wise enough to find areas where we can successfully intervene.

So, let me summarize ten points that I have discussed in this book:

1. The crowd is formed through non-linear human interactions, which may be significantly more complex than we believe.

The lesson: We usually think of the crowd as a mob moving along a simple and predictable trajectory. Sometimes, this simple description is good enough to guide our conduct. However, to gain an edge as individuals, we should focus on the more complex aspects of crowd behavior. For example, a crowd that behaves as an unstable system may suddenly switch to a different regime of behavior. The exact timing of this shift is unknown. Understanding the complexity underlying this regime shift, the individual may adopt a more cautious approach that takes into account the unpredictability of the exact shift.

2. Understanding the constraints operating on the system (e.g., bounded exponential growth) is important for betting against the crowd.

The lesson: The idea of constraints compels us to think about the bounded behavior of the crowd and its limit lines. For example, the gold rush on the stock market, emerging financial bubbles, and Messianic political optimism are all forms of crowd behavior that are blind to the constraints operating on the system. A specific and current example is the American rhetoric seeking the democratization of cultures constrained by worldviews that have nothing to do with democracy or constitutional republics.

3. Learned ignorance should guide us, at least as the first step.

The lesson: Learned ignorance is an approach requiring us to acknowledge that our knowledge is limited. For example, a critical thinker would have asked the Israeli PM whether he knew if and when Hamas might turn its plans into a vicious operation to invade Israel. The PM could have answered by saying, "We believe that ..." or "The probability is ..." but these answers do not indicate knowledge. Answering "I don't know" would have led to a different strategy that could have prevented the atrocities, including the slaughter of babies, the rape and murder of young women, and the kidnap and murder of senior citizens. Adopting the idea of *docta ignorantia* forces us to avoid the biases of narcissism and self-delusion. In too many cases, we simply don't know. Understanding a complex crowd should start with this approach.

4. We should adopt scientific thinking in the absence of truth.

The lesson: As explained by von Neumann, science is about modeling. Our models are tools that aim to represent reality in an informative way. Instead of seeking the "truth" underlying the crowd's behavior, we should focus on modeling the behavior using measurable facts. The idea of profound simplicity urges us to model the underlying dynamics of a crowd using suitably simple models and focusing on facts rather than on an illusion of truth that may simply drag us back into an old religious rhetoric rather than guiding us toward the least stupid action.

5. Identifying informative signs (e.g., the shoeshine boy) may be crucial for a contrarian approach.

The lesson: The art of modeling is closely associated with the identification of informative signs. Building a simple model of the crowd may lead us to a few informative signs that are more than enough to guide our behavior. A model is a simple and powerful representation of a system. It includes a few measurable variables. These are "facts". Facts can be interpreted as highly important signs for betting against the crowd. For example, I measured the ungrounded optimism of football fans. This measurable fact is a sign of a fixed belief. Any individual who understands the wishful thinking of these fans can easily bet against them.

6. Uncertainty is both a problem and a resource.

The lesson: We may naively think of uncertainty as a problem, but uncertainty is both a problem and a resource. Misunderstanding uncertainty is a problem. I explained this point in Chap. 9, where I analyzed the complexities of armed conflicts. We must understand uncertainty, be aware of uncertainty, and manage uncertainty as much as we can. Uncertainty is also a resource when betting against the crowd. For example, when the crowd experiences a state of chaos, we may have the opportunity to channel its behavior. As I explained elsewhere, Russian interference in the American elections aimed to increase chaos. The logic probably underlying this move is that a state of instability may be better used to influence the American public and channel them into the "right" direction.

7. Identifying short-time frames of intervention may give us an edge.

The lesson: Betting against the crowd requires a deep understanding of scaling. Both size and time scales. In some contexts, the best prediction exists for short time scales where constraints are known and can be used to gain an edge. While strategic and long-term thinking is critical, I embraced the idea of winning in very short time frames. This direction does not exclude the possibility that different time frames are required for intervention in other contexts. Sensitivity to time and size scales is therefore highly important for understanding crowds and betting against them.

8. The amount of required information is determined by the exponential decay function.

The lesson: How much information is enough? How much information is required to reach a good enough decision? How much information is required to bet against the crowd? I answered these questions by proposing a different approach. Instead of focusing on the information, I focused on the dynamics of loss and pointed to the exponential decay of information. This general function, which has its contextual particularities, determines a window of opportunity to collect information and take action. Understanding this dynamic of decay is crucial in choosing the best thing to do.

9. Extrapolation of a trend may teach us about our ignorance more than about the future.

The lesson: Like Russell's poor hen, we are extrapolating creatures seeking to predict future trends on the basis of our experience. Understanding prediction and extrapolation errors may draw our attention to the limits of our understanding and prediction and mark the area where dragons live.

10. We all live in Plato's cave, whether we like it or not. We must adopt a creative approach to transcend the limits of understanding.

The lesson: We observe low-dimensional projections of a highly complex reality. While some modern and sophisticated technologies are capable of representing and using high-dimensional spaces, it is difficult to understand them and trust that their measurements represent all the relevant features at the relevant time. Using low-dimensional representations for working in a complex reality is a challenge. For example, the behavior of a crowd involves non-linear interactions between its members. In most cases, we cannot measure these interactions, and not usually in real-time. However, we can identify some low-dimensional signals and exploit them to "reconstruct" a higher-dimensional space that we can use. For example, we may use a low-dimensional signal, such as the level of domestic violence, and follow the measurements of the signal over time. Using different techniques for measuring the minimal higher-dimensional space in which this signal "lives," we may track the evolution of its complexity and learn something interesting about the society in which it has been generated.

A Final Word

My late grandmother used to say in Yiddish, the Jewish dialect, "Mentsh Trakht un Got Lakht," meaning "Man plans, and God laughs." This proverb emphasizes the unpredictability and uncertainty of life. In contrast with Tolkien's hobbits, who lived a stable and peaceful life in a remake of pastoral England, my grandmother experienced life's uncertainty and the crowd's ultimate madness. Individuals and crowds face uncertainty, but in this book, I repeatedly emphasize that the individual within the crowd may avoid and even exploit some uncertainty to bet against the crowd. This is not the usual case. Our control as individuals is minimal, and we may not often find pockets of opportunity where we can act against the crowd. Despite these difficulties, it is an ideal we must strive for as individuals.

The ten lessons listed above are some of the most general ones that can be learned from the book. The specific examples and experiments I have presented aim to deepen and illustrate these general ideas. There is nothing to be added, and the reader has hopefully learned something meaningful, if not as a guide for the individual rebel, at least as a road map for avoiding our stupidity within a crowd.

Reference

Neusner, J. (Ed.). (1974). *The Mishnah: A new translation*. Yale University Press.

Author Index

Y. Neuman, *Betting Against the Crowd*, https://doi.org/10.1007/978-3-031-52019-8

Subject Index

Y. Neuman, *Betting Against the Crowd*, https://doi.org/10.1007/978-3-031-52019-8

Printed in the United States
by Baker & Taylor Publisher Services